METHODS OF SCIENCE

Methods of Science

An Introduction to Measuring and Testing
for Laymen and Students

by

E. L. DELLOW

UNIVERSE BOOKS : NEW YORK

Published in the United States of America in 1970
by Universe Books
381 Park Avenue South, New York, N.Y. 10016

Library of Congress Catalog Card Number: 71–121793
ISBN 0–87663–129–4

Printed in Great Britain

Contents

Illustrations

Illustrations

Preface

One of the paradoxes of our time is that, while we live in an increasingly scientific and technological society, the ordinary citizen's knowledge of and interest in scientific and technical matters seems to be declining. To anyone who has himself experienced the fascination and excitement of these subjects, this must inevitably be a matter for regret; to those who are charged with the education and training of the next generation of the engineers and scientists upon whom we are all now so heavily dependent, it is increasingly becoming a matter of alarm.

If this lack of interest, which sometimes amounts almost to hostility, were combined with a total rejection of the scientists' work, it might (in my opinion, it would) be entirely regrettable, but it would then at least be logically consistent. This, however, is not the present trend. On the contrary, the fruits of science are desired more than ever; only the will to cultivate the orchard is lacking. The ultimate results should be plain enough. On the international scale, leadership, possibly even direct control of affairs, will pass to those nations who are prepared to cultivate their science and technology. On the national scale, the continuance of the trend would lead first to the emergence of an esoteric 'priesthood' who alone would be fitted to take many of the decisions of modern life, and ultimately to complete Technocracy. Such a thought is anathema to many. The only answer—the only defence, if you care to put it that way—is knowledge.

To convince the reader of this is one thing; to show him the way to the knowledge is quite another. During the earlier part of the twentieth century, until shortly before the Second World War, a band of brilliant scientists who were also brilliant writers raised the popular exposition of science almost to the level of an art form. It is a matter

9

of profound regret that, with a few notable exceptions, their work has not been carried on. On the other hand, any book which even appears to invite comparison with the work of these masters necessarily starts under a heavy handicap. As excuses for my temerity, I can offer only the fact that my aim is the same as theirs—to show that science and technology are interesting matters, worthy of study; and a belief, based on the considerations set out above, that the need is urgent.

The purpose of this book, then, is not new. It is simply to arouse the layman's interest in scientific and technical matters. If I can claim any originality at all, it lies in the method adopted to achieve this aim. Quite simply, the attempt is made to exploit a trait inherent in most of us—a certain delight in watching other people work. Rather than setting out to dazzle the reader by a recital of the achievements of science and technology, I have tried to show how the knowledge, which alone makes the spectacular achievements possible, is gained, and also to give some insight into the ways in which scientists and technicians think. Of course, if the reader also manages more or less painlessly to acquire a grasp of some scientific facts and principles, that is all to the good.

The plan of the work is fairly straightforward. Most of the really heavy going, for the person who has not previously taken much interest in these subjects, will be found in the first three chapters, which contain all the 'basic' material necessary to an understanding of the rest. While much of the remainder is quite intelligible without this foundation, I should always recommend a reader new to a subject to begin at the beginning. However, this is a matter of temperament: if the reader feels inclined to start in the middle because something there interests him, and to return to the beginning afterwards when the need arises, by all means let him do so.

The remaining chapters form a brief, reasonably simple, and necessarily highly selective guide to some of the techniques of research, discovery, and invention. It is not pretended that the topics dealt with are necessarily the most important ones. They have rather been chosen as being likely to be of general interest. It is too much to hope that in making a small selection from an enormous field, my own pet interests will not have influenced the choice; nevertheless, I hope that the selection is sufficiently varied to appeal to a good many people. Though the approach is not deliberately historical, a fair amount of historical detail is included, since this, rather than a rigidly logical development from first principles, often enables the non-specialist reader to orientate himself better in a new subject. No one will ever acquire a real feeling for scientific and technical things simply by reading about them, so it is suggested from time to time that the reader should perform a simple experiment. The ones described should be well within the capacity of even the most thumb-fingered individuals, and require no apparatus or materials which cannot be found in most homes.

One of the pleasures denied to a writer of a book such as this is to be able to claim that such-and-such 'is completely original' or 'has not previously been published elsewhere'. Much of the information in this book was gathered in the course of professional activities which have brought me into fairly close contact with scientists and technical men, and this seems the appropriate place to acknowledge my indebtedness to all those scientists, engineers, writers, editors, and professional colleagues who, to a greater or lesser degree, and very often quite unconsciously, have helped me to acquire that knowledge.

Certain specific acknowledgements are made in the appropriate places, but special thanks are due to my wife, both for her support and understanding during the long process of preparing and writing a book in the spare time which should by rights have been spent by the domestic hearth, and also for nobly agreeing to act as a 'guinea pig' by reading the manuscript to see whether my exposition was as clear as it should be. Any remaining faults or obscurities are all my own work.

E. L. Dellow

1

The Scientific Method

Human beings come in a wide variety of shapes, sizes and colours, and have widely differing personal characteristics —in fact, no two are alike. However, they must have some attributes in common, and many attempts have been made to sum up those attributes in a few words. Such exercises will not be attempted here; it is enough to point out that there is one attribute that every normal human being possesses to a degree far greater than any other animal, and that is an overwhelming curiosity about everyone and everything. Above all, we want to know. And if we don't know, we want to find out.

Children want to find out about everything. As they grow up, they realise that life is too short for the fulfilment of this laudable ambition, and that in order to do many of the other things which they want to do, or have to do, they must confine their curiosity to more or less narrow channels. Most of us have accomplished this fairly successfully, and we are accustomed to say that we are 'interested' or 'not interested' in this or that—though given the chance, our inborn inquisitiveness will generally pop up again.

The ways and means of satisfying curiosity—of finding out—are many and varied. Anyone who has spent much time in the company of young children knows about one of them—asking questions. Another exclusively human attribute is the ability, once having acquired knowledge, to retain it and pass it on to others—from generation to generation, if need be. Many of our methods of finding out are simply more or less sophisticated versions of the child's method of asking questions. The torture chamber and the library have this much in common—they are both devices for tapping someone else's knowledge.

However, it may happen—it very frequently does happen—that the information which we require is not possessed by anyone else, and we are thrown upon our own resources. We must ask our questions of the universe in which we live. There are various ways of doing this. Early men did it, as their descendants still very often do, unconsciously. They learned by experience. Intellectual effort and mystical experience are other methods which have been, and are still, tried, and all of these ways of finding out are as old as the human race. There is one method, however, which though it is comparatively new, is supremely successful. In the four centuries or so during which it has been properly understood and used, it has not only enlarged Man's stock of knowledge many thousandfold, but has helped to bring about changes in his outlook and, indeed, in his physical environment, far outstripping anything which has happened in the rest of recorded human history. This is the method of scientific inquiry.

This method, whether he like it or not, whether he realise it or not, impinges upon and affects the daily life of anyone living in a civilised society—it is far from being the sole concern of white-coated gentlemen in universities and the 'back rooms' of industry.

SCIENCE AND TECHNOLOGY

Before we go any further, it will be as well to define a few terms. *Science* is the body of organised knowledge, which has been obtained, and is constantly being added to, by means of scientific method. More loosely, the pursuit of new knowledge by this method is also termed 'science'. *Pure science* is knowledge, or the quest for it, for its own sake. Reduced to the starkest terms, it is nothing more nor less than the frank, though disciplined, exercise of curiosity.

Applied science involves the use of knowledge so gained for practical purposes, generally in industry. *Technology* is the application of scientific principles to technical processes, and the term can also be applied to the body of basic principles upon which such processes rest. This word is often used rather loosely. Though technology and applied science do overlap to a considerable degree, they are not quite the same thing. Perhaps a couple of examples will help to make the distinction clear.

In 1856 the English chemist, W. H. Perkin, discovered a hitherto unknown substance in the course of some experiments with aniline, which is a chemical derived from coal tar. Further experiment showed that this substance, afterwards christened 'mauve', was a very efficient dye for silk. Perkin was a brilliant chemist, but he also possessed what is by no means universal in scientists, a well-developed business sense. He immediately patented his discovery, and set up a factory to manufacture the new dye. This venture was so successful that by the time he was thirty-five (he was only eighteen at the time of the discovery) he was able to retire from business and devote the rest of his life to research. This is an excellent example of applied science; in fact it represented the beginning of a new industry, though this was not realised at the time.

A good example of technology is the improvement which took place in steelmaking during the latter half of the nineteenth century. The importance of this to engineering, and to industry generally, can hardly be exaggerated. Up till that time, progress had been comparatively slow: developments in the iron and steel trade had nearly all been the result of trial and error, combined very often with a large element of luck; usually the discoverer of a new or improved process tried to keep it to himself. With the greatly increased demand for iron and steel brought about by the continuous expansion of engineering, it was realised that these somewhat haphazard methods were not enough. Innovators grasped the fact that it was no use trying to improve a process if, as was usually the case, they did not fully understand why it worked in the first place. They therefore began to study the existing processes, and eventually arrived at an understanding of the chemical and physical bases of them. Improvements could then be suggested on a basis of knowledge rather than guesswork, and it is not surprising that they were usually successful and sometimes revolutionary.

Thus technology implies research and a scientific way of thinking, but the research need not necessarily be carried out in a laboratory, and knowledge is gained, not for its own sake, but as a means to an end. It is wrong to use the words 'technology' and 'technological' if this element of basic understanding is not present. Many people seem to regard 'technological' simply as an elegant variation of 'technical', but this is not so. A *technician* is one who can operate a process, though nowadays he must have more than a nodding acquaintance with its technology if he is to be regarded as fully efficient. A *technologist* may not be able to go into the workshop and do a job himself (though frequently he can), but he does either understand the underlying principles, or is capable, by reason of his training, of applying scientific method to discover them, and so make improvements in the process or suggest reasons why it is not going well.

The point should perhaps be made that the rise of technology has in no way diminished the importance of pure science. All knowledge is potentially useful, and time spent in adding to the total stock of human knowledge can never be wasted. The present writer once had the privilege of being shown an important new process by one of its inventors. One of the techniques which was essential to the success of the process had not been worked out from scratch, but had been found in a comparatively obscure paper describing the results of a piece of 'pure' research, and published some thirty years earlier. This explains why many large companies consider it worth while to maintain research departments which, in addition to their more obvious functions, undertake a considerable amount of 'pure' research. This is not done in the expectation that the scientists will stumble upon a commercially valuable discovery (though that hope is perhaps not entirely absent), but because it is recognised that it is ultimately to everybody's benefit that they shall be enabled to continue making discoveries of any kind.

DESCRIPTION OF THE SCIENTIFIC METHOD

'Scientific method' has been mentioned several times above. What is this? Essentially, it is a set of rules for thinking, and a way of looking at the world. It has been said that science is organised common sense. It has also been said that 'there's nothing so uncommon as common sense'. There is no particular reason why this method of thinking should be the prerogative of trained scientists, but the fact is that most people do not organise their thought-processes in a way most likely to lead to results. The 'classical' description of the scientific method, as it was taught to the present writer, and as it is still taught to every generation of students, is as follows.

The process begins with the *collection of data*. Data for scientific and technical purposes are obtained in the first place in one of two ways, by *observation* and by *experiment*. The method of observation appears simple enough: we watch carefully and record what we see. However, this is not quite as easy as it sounds. Anyone who has had occasion to listen to court proceedings arising out of a road accident may be forgiven for wondering, sometimes, whether the witnesses are all describing the same occurrence. It not infrequently happens that cross-examination reveals not only that a witness failed to see something which happened, but that he saw, or thought he saw, something which most definitely did not happen.

Most people look, but do not see. Readers of detective fiction will recall how Sherlock Holmes was enabled to perform apparently miraculous feats of deduction simply because he had trained himself to notice things that his contemporaries missed. These stories are by no means exaggerated—the fictitious character of Holmes was based upon a real person—and scientific observers must cultivate something of the same ability. This by itself, however, is not enough. Psychology was an infant science when Conan Doyle wrote his stories, but we now know a good deal more about *perception*, and two undoubted facts which have a bearing on the matter of observation are first, that emotional factors play a considerable part in perception, and second, that the human mind (or brain) has a distinct tendency to impose order on the outside world—an order which may not in fact be there.

Accurate and reliable observation therefore requires much training and practice, and relatively few people are really good at it. The tendency is therefore to try, so far as possible, to replace the human observer by machines and instruments, whose impersonal records can be examined at leisure and, if need be, checked and analysed by more than one worker. Many examples will be found in this book. Incidentally, this fallibility of the human observer explains why professional scientists display so much caution (to put it mildly) over reports of such things as 'flying saucers'. It is not that they have a snobbish outlook towards amateurs, as so many people seem to believe, but simply that they know the difficulties involved, and are reluctant to place much confidence in the reports of untrained observers who, very often, are only too obviously quite unaware of these difficulties.

The Scientific Method

In their early stages most sciences are primarily observational, and there are a few quite highly developed sciences which, by their nature, remain so. Good examples are astronomy and anthropology (the study of man as an animal). Other scientific studies, mainly in the biological field, also retain a strong element of observation. It is perhaps worth remarking in passing that amateur workers, provided they are properly trained in observational technique, can and do still make important contributions to these sciences. This is scarcely possible nowadays in the more advanced physical sciences, where the experimental method is pre-eminent.

Experimental work

Experiment is the second method of obtaining data mentioned above. In many ways, the method of *controlled, repeatable experiment* is the very essence of the scientific method, and it certainly distinguishes scientific procedure from other methods of obtaining information. It will, therefore, be worth while to look at it in some detail.

Observation of the world as it is forms an excellent, and indeed indispensable, method of gathering information, but the time inevitably comes when it is no longer sufficient. Some phenomena are extremely rare in nature; one might spend a lifetime of observation and witness them only once or twice—not a very efficient way of learning. Or it might simply be that conditions of observation are excessively uncomfortable or inconvenient. It is quite possible, and for some purposes absolutely essential, to go and observe, say, polar bears in their natural surroundings. This method commends itself only to the hardy few. It is also possible to learn a lot about polar bears in more comfortable surroundings: we capture a few of them and put them in a zoo. In so doing, we create an artificial situation to suit our own purposes, and this is the first step towards the experimental method.

The aim of a properly conducted experiment is to create an artificial, reproducible situation in which the factor to be studied can be isolated and observed. This is not quite as simple as it sounds, and the design of experiments is a highly specialised business. The first care of the experimenter must be to ensure that he has in fact set up a truly reproducible situation. This is not primarily for his own convenience—though obviously an experimental set-up which does not behave the same way tomorrow as it does today is not much use to anyone—but because, if the results are to have any validity, they must be capable of being reproduced by other workers using the same set-up perhaps on the other side of the world. This is the meaning of repeatable experiment—its importance cannot be overstressed.

The second essential is to isolate the factor under study. If this is not done, it is quite possible, even likely, that fallacious conclusions will be reached. This can be a very complex and difficult matter, but perhaps the principle can be explained with the aid of a simple, indeed a trivial, example. Suppose you drop a lump of sugar into a cup of tea, and stir it. When you taste the tea, it is sweet. What was the cause of the sweetening? Obviously, the sugar. Obviously, that is, until some awkward individual comes along and asks 'How do you know it was the sugar and not the stirring?'

With the experiment as described, you do not know. It is necessary to separate the two factors. It is easy enough to show that stirring a cup of tea without adding sugar does not cause the tea to become sweet, and this seems to settle the matter, until it is noticed that simply adding a lump of sugar to an unstirred cup of tea does not seem to affect its sweetness, either. To be certain, it is necessary to take two identical cups of tea, add sugar to one, and stir both. Only the one with the sugar becomes sweet, so it is a reasonable inference that the sugar is the cause of the sweetness. This can be confirmed by again taking identical cups of tea, adding sugar to one, and allowing them to stand. After a time the sugar diffuses through the tea, and sweetens it, so the final conclusion is that the sugar causes the sweetness, and the stirring merely hastens its dissolution.

This is a controlled experiment. The factors which might have affected the outcome have been isolated and studied separately. The cup without sugar in these experiments is described as a *control*, and it is always advisable, if it is at all possible, to arrange for a control or 'blank' experiment as a check. In any case, only one thing should be studied at a time, and if there are several factors which can be altered, observations and measurements on one of them must be completed while the others are held constant; only then may another factor be selected for alteration, the same precautions being observed. (It should, perhaps, be stated for the sake of completeness that there are methods whereby valid results may be obtained when observing several variable factors at once, but these are advanced techniques, well outside the scope of this book. In any case, no scientist adopts such methods unless the simpler ones already described are, for some reason, out of the question.)

Systematisation

The next step in the scientific method is *systematisation of data*. A simple collection of facts may be interesting, but it is not of much use to anyone. A primary aim of science is *unification*. If we can show that two apparently unrelated facts are actually connected, we have taken a great step forward in our effort to understand them. Hence the importance of systematisation, ie getting data into coherent form and order. This may be merely a matter of careful arrangement and tabulation, or the data may have to be subjected to some mathematical manipulation, as outlined in the next chapter. It may even be necessary to make a selection, though this is to be avoided if possible, and conclusions drawn from selected data are usually regarded with some suspicion. In any case, if the worker does make a selection, it is a cardinal principle, and a matter of absolute honour, that he shall make the whole of his data, not only the selected items, available to other workers, so that his conclusions may be subjected to critical examination.

It will already be obvious that the scientific view of the universe takes it for granted that there is an underlying order and pattern in everything. This view has been challenged; it has been suggested that the order exists only in the human mind, which imposes it upon the actually chaotic universe. This is a philosophical question; for practical purposes, so long as there is order (as there is) it is immaterial *where* it exists. The working scientist (this term, incidentally, can be taken to include technical men, even where this is not specifically stated), finds that it simplifies his task if he assumes order to be present in the universe. This assumption is obviously important in the systematisation stage; it is vital in the next steps.

Hypothesis

So far, we have imagined workers in the field and the laboratory collecting data and arranging them systematically. The next task is to show how the new facts fit into the existing body of knowledge, and whether, and if so how, they are themselves connected. This is, of course, an intellectual exercise, and nowadays it is very often undertaken by separate workers. Formerly it was quite usual for a single scientist to make observations, conduct experiments in his own laboratory, classify his results, draw inferences from them, and publish the results of his deliberations to the world. Nowadays, the body of scientific knowledge has grown too large and complex for this, and just as, about a hundred and fifty years ago, the 'natural philosopher' who took the whole of science for his province gave place to specialists in physics, chemistry, geology, etc, and at the beginning of this century even these sciences began to be subdivided, so the tendency is now for scientific workers to specialise as 'experimenters' or 'theorists'.

This division has gone farther in some disciplines than in others, and is still absent in a few. It is not without its own dangers; nevertheless, it becomes more and more necessary if any effective work is to be done.

Whoever actually does it, sooner or later an explanation of the facts is evolved, and this is termed an *hypothesis*. It must fit the facts, all the facts, or it is useless. If it does fit the facts, it can be tested. The test of an hypothesis—and the only test which is accepted as valid for scientific purposes—is *prediction*. Thus, if a series of experiments, conducted with the precautions described above, has yielded a number of measurements for a certain factor under differing, known conditions, it may be possible to set up an hypothesis linking the value of the measurements with the varying conditions. For example, it is noted that the pressure of steam in a boiler varies with the temperature of the water, and it is found that the relationship can be expressed by a mathematical formula. The hypothesis in this case is that pressure and temperature are directly linked in a manner described by the formula. This is easy enough to test. If it is true, it should be possible to say what the pressure will be for a given temperature or temperatures which were not included in the original experiment.

A prediction of this kind can, of course, only be tested by another experiment, a fact which accounts for the central importance of experiment in the scientific method. If the new experiment does not confirm the prediction, and assuming that no fault can be found with the experimental method, then the hypothesis cannot stand. It must be modified to take account of the new result, if this is possible, or discarded altogether. To discard an hypothesis which may have been the result of considerable intellectual effort, and with which its creator may well have become to some extent emotionally identified, is something which calls for considerable strength of mind and intellectual honesty. Nevertheless, the rules of science allow of no exceptions, and it says much for the integrity and character of its practitioners that they are seldom slow to retract an hypothesis which is shown to be untenable—a refreshing contrast to the behaviour of politicians placed in a similar position! If, however, repeated experiments show that the predictions are consistently verified, the hypothesis is elevated to the dignity of a *Law*, and it can be used as a foundation for further work.

It will be evident that this is a self-perpetuating process. The experiments required to test an hypothesis may themselves reveal fresh facts, which provide the starting point for further work. Thus the progress of science is continuous. When a series of hypotheses has been built up, and these are found to be consistent with one another and with the observed phenomena, the result is a scientific theory. Because of its wider scope, a theory is more important than a simple hypothesis; it is also more difficult to test, but it is susceptible of exactly the same test of prediction. Predictions from theory are apt to be more spectacular than those from simple hypotheses; one such, which was certainly vindicated in a somewhat emphatic fashion, was the prediction that in certain circumstances a nuclear 'chain reaction' was possible.

AN EXAMPLE

It will perhaps help to clarify the above discussion if a 'classical' example of the scientific method is given. Unfortunately, it must be admitted that there is a good deal of doubt as to whether the experiment which will be described was ever actually performed, but for the stated purpose, this does not really matter. Most of the story, which is a well-known one, is verifiable.

The Scientific Method

One of the founders of what is now the United States of America was Benjamin Franklin. Like many of the intellectual leaders of his day, he had a colourful and varied career, and was far from being a narrow specialist. Among other things, he studied electricity, which during the latter years of the eighteenth century, was just beginning to engage the serious attention of scientists. Experiment had shown (and Franklin is known to have performed all the relevant experiments himself) that bodies could be electrically 'charged'; that the charges were of two distinct kinds, that under the appropriate conditions a 'discharge' could take place between two oppositely charged bodies or between a charged body and the earth, and that such a discharge was accompanied by a spark—the heavier the charge the larger the spark—the bodies then reverting to their normal uncharged state.

Certain parts of North America are subject to frequent and violent thunderstorms, and Franklin took a great interest in the phenomena of thunder and lightning. He observed, among other things, that lightning passed between cloud and cloud, or between cloud and earth. Upon systematising his observations and experimental results, he was led to propound what was for that time a somewhat startling hypothesis: that lightning is an electrical phenomenon; in fact, nothing more nor less than a giant spark.

It will be seen that this hypothesis fits the facts very well. We have the discharge between charged bodies (the thunderclouds) or between the cloud and earth. People had been 'struck' by lightning, and it was known that a sufficiently large electrical discharge could kill at least an animal, and could stun a man. It had to be assumed that the charges involved in lightning were far greater than anything that had been produced in the laboratory up to that time, but this was perfectly logical; a thundercloud was far larger than the laboratory equipment.

The hypothesis is also open to the test of prediction. It was known that electrical charges could be drawn from a charged body by means of a conducting material—metal rods and chains were used in the early days—and Franklin predicted that it would be possible to draw a charge from a thundercloud in a similar manner. The story goes that he put the matter to the test himself, by flying a kite into a thundercloud. When the string of the kite had become wet, it conducted electricity, and Franklin was able to draw sparks from a key tied to the string.

As mentioned above, this story is open to considerable doubt. If the experiment was performed as described, it was a courageous, not to say foolhardy one, for Franklin ought to have known well the forces he was dealing with. If he actually flew a kite into a thunderstorm, he was lucky to have escaped with his life, and a Russian professor was in fact killed while trying to repeat the experiment. It is significant that Franklin's own autobiography gives no direct description of the event. However, it is really immaterial whether or not Franklin actually performed the experiment himself, for he had communicated his ideas to a French colleague, D'Alibard, who did carry out a successful experiment to test them (with rather more safety precautions) in May 1752.

Thus the scientific process had been carried through completely, from experiment and observation through systematisation to hypothesis and prediction, and once more back to experiment as the final arbiter. The matter had been proved: lightning was an electrical phenomenon. Incidentally, a practical application of this piece of pure science came almost at once, in the form of the lightning conductor.

PSYCHOLOGICAL CONSIDERATIONS

Some of the precautions necessary in experimental work have been outlined already, but one other matter deserves mention; this is the human factor. It will be obvious by this time that an experiment is seldom or never simply an open-minded attempt to gain information—any information. No one nowadays performs an experiment 'just to see what happens', and indeed this would be likely to be a rather profitless proceeding. An experiment is nearly always carried out to test an hypothesis, or because its desirability has been suggested by the results of earlier work.

It follows that the experimenter, unless he is capable of an almost superhuman detachment, will want his experiment to be successful, in the sense of demonstrating the correctness of his ideas. Strictly speaking, there is no such thing as an unsuccessful experiment—all experiments show something, even if only that an idea is untenable.

The imperfections of the human observer, considered as an instrument, have already been pointed out. Time after time, and not only in scientific contexts, it has been shown that faulty observations did not just occur at random, but were conditioned by the observer's preconceptions and beliefs. In other words, it is fatally easy to see, not what happens, but what one wants to see.

A scientist or a technician must be aware of this tendency, and must guard against it as far as he can. Some of the safeguards—the increased use of instruments, observations by more than one worker —have been described. The separation of experimentalists and theoreticians, also alluded to above, is a further safeguard. This separation has yet another advantage: the talents required for delicate and precise manipulative work, and careful observation, are by no means the same as those involved in working out mathematical relationships and constructing theories in the quiet of the library. We all know, from everyday experience, that people are not only happier doing work they like and are temperamentally fitted for, but the results also are usually better.

Of course, there have been, as there still are, many scientists who are equally good at theoretical and experimental work—had it been otherwise, science would probably never have developed. But the day of the all-rounder and the individual worker is passing. Science, and particularly technology, is becoming a matter of teamwork. The occasional genius does not fit into this pattern—he never did. When he comes along, he must follow his own bent—to the ultimate benefit of everyone. But geniuses occur perhaps once in a generation, and the world is now so dependent upon science and technology that we can no longer afford to wait for them to be born. People of lesser ability must do their best to keep progress going.

One consequence of this is that the scientific method, as described, really does tend to be the process by which discoveries are made. It has always been the ideal, yet, if the history of science is closely studied, it can be seen that the sudden intuition, the brilliant leap in the dark, or the plain old-fashioned (but unadmitted) guess have been quite as common as the more logical and ordered process. It would be idle to pretend that these processes play no part in scientific discovery today; they do, especially in the hands of the odd genius mentioned above. In a sense, what is involved is the ability to perceive a pattern, and it is the mark of the brilliant innovator in science that he sees the pattern that nobody else has noticed.

The formal scientific method is a set of rules for enabling ordinary people to do what has hitherto been the prerogative of genius. Much of the foregoing has been written as though new discovery were the sole aim of scientific work, but while this aspect is extremely important, many investigations, especially in industrial laboratories, are undertaken for much more mundane purposes. This kind of work may be divided, broadly speaking, into two categories: first, the solution of specific problems, for example finding why a process is not working as it should, making improvements in a process which is going well but which might be better, or ascertaining the best way of making or doing something. The second category is usually referred to as 'quality control' which generally means some kind of testing.

A good deal of this is therefore either direct technology or has technical applications, but it requires the scientific method. It is this kind of scientific work which is the subject of most of the rest of this book, and although 'science' is often written by itself, the words 'and technology' can usually be taken as read. The idea, propounded above, of science as a method of asking questions of the universe, remains valid in this context, but we shall be more concerned with the questions, and with the ways of putting them, than with the answers. These, where they are already known, are recorded in the literature of science and technology, where the interested inquirer may find them.

2

Adding up the Answer

Measurement is of the essence of the scientific method. This is not always immediately obvious, and indeed, as recently as fifty years ago, it was still possible to do important and original work in some branches of science—mainly in the biological sciences—without making a single measurement. The situation has changed drastically today, and it is now quite clear to all scientists, as it had long been to the physical scientists, that quantitative work is by far the most important aspect of the scientific method.

Without measurement, the physical sciences would not have progressed very far, and technology could scarcely have begun. A vast amount of the information found by or required for scientific and technological procedures is in the form of weights, measurements, or instrument readings of some kind. If, as suggested earlier, we regard the scientific method as a way of asking questions about the universe, then it is clear that the answers to many of our questions will be numbers.

Now the branch of knowledge which deals with numbers (in the broadest sense of the word) is mathematics, and some knowledge of mathematics is essential to any real understanding of science in general. At this stage I can well imagine many readers putting the book down with the remark 'This is not for me—I never was any good at mathematics, and I never shall be'. But wait a moment! When did you last allow a shop-assistant to get away with giving you short change? How do you know how much fuel your car will need for a given journey, or when to start the journey in order to arrive at a given time? If you are interested in sport, do you follow the cricket or baseball averages, or racing odds?

The fact is, of course, that most of us are accustomed to perform, every day of our lives, mathematical operations which would have utterly dismayed our not-very-remote ancestors. In a modern technological civilisation, the wholly 'innumerate' person is as rare as the wholly illiterate one. It is true that the level of 'numeracy' is not, in general, as high as that of literacy, but this is in the main due to a totally irrational fear of mathematics, largely the result of unimaginative teaching, but partly also stemming from a misunderstanding of the degree of knowledge and effort required.

It is not necessary to the understanding of a mathematical argument that one should be capable of formulating it for oneself; neither is it essential to know how some piece of information, presented in mathematical terms, was obtained before being able to use it. Naturally, the more mathematics one knows, the easier it is to follow an argument presented in mathematical terms, and the deeper one's insight. But to start with, at least, no vast knowledge is necessary. Of course, there are people who take to mathematics without apparent effort, just as there are people who can master a dozen languages while most of us think we have done well to acquire a reasonable knowledge of two or three. At the other end of the scale, it is nowadays recognised that a few unfortunates are illiterate not because of any lack of native intelligence, but because they are 'word blind'—a pathological condition requiring special treatment. There is not the slightest reason to suppose that a total inability to understand mathematics is any more common than word-blindness, so, assuming the reader to be possessed of his share of ordinary common sense, let us to business.

It will be obvious that in a single chapter of a comparatively short book, it is quite out of the question to present anything in the nature of a comprehensive course of mathematics. Numerous excellent books have been written which aim to dispel the irrational fears mentioned above, and present mathematical knowledge in a more-or-less 'palatable' form. Some of them are mentioned in the Reading List at the end of this book. Here, it will be assumed that the reader has at some time completed at least an elementary school course of mathematics, even though he may since have forgotten much of it, and the attempt will be made to show how mathematics is *used* for scientific and technical purposes. This, in the present writer's opinion, is the key to the whole business: it is unfortunately true that in far too many schools, even today, mathematics is taught as though it were a purely intellectual discipline, like the 'dead' languages, totally divorced from the real world.

THE NATURE OF MATHEMATICS

Many definitions of mathematics have been written from time to time, but for the present purpose we shall be justified in regarding it as having two major functions, though these overlap to a large extent. First, then, it is a set of rules for carrying out certain mental operations, primarily (though not quite exclusively) upon 'numbers' in the broadest sense. Notice two things in this connection. The rules are justified because they have been shown, by experience over the years, to lead to repeatable and meaningful results if correctly applied, but many of them were in use long before anyone understood *why* they gave the results—in fact there are some mathematical procedures now in use of which the theoretical basis remains uncertain. Further, in some cases at least, we can change the rules and get different results, which are equally repeatable. Whether or not they are meaningful depends upon one's point of view. For scientific and technical purposes, we regard them as meaningful and 'valid' if they correspond to something in the physical world, although a 'pure' mathematician would not accept any such limitation.

Second, mathematics is a system of symbolism, by means of which we can express ideas for which ordinary language is either impossibly clumsy or has no words at all—if you like, a system of shorthand. Thus, '7' stands for 'seven'—not very impressive, though it saves four characters; 'n' stands (usually) for 'any number'—rather more saving; 'π' stands for 'that number which represents the ratio of the circumference of a circle to its diameter'—a very considerable saving in time and space; while the number usually denoted by 'e' cannot really be explained in ordinary English at all. To say that it is 'the basis of the Napierian logarithms' is no explanation unless one knows what Napierian logarithms are—the explanation can only be made in mathematical terms, though in those terms it is simple enough.

Mathematics also embraces the subjects of geometry and trigonometry, which are respectively the study of the properties of space and that particular branch of it which deals with the special proper-

29

ties of triangles. These are important subjects technically, since they form the basis of surveying and navigation, to name only two applications. It might be thought that these subjects do not fit into the definition of mathematics given above, but it can be shown (and it is shown, in advanced textbooks) that all the statements and proofs of geometry and trigonometry can be reduced to symbolic form, and it is sometimes more convenient to handle them in that form.

One of the most important aims of mathematicians is *generalisation*. The simplest way of setting out an accurate right-angle without elaborate equipment is to draw a triangle with sides 3, 4, and 5 units (inches or centimetres, for example) long. The angle opposite the longest side is a true right-angle. This fact was known to the ancient Egyptians before 4000 BC and is still used every day in all kinds of practical work. Experiment shows that it is possible to construct right-angled triangles with varying lengths of side, but very few of them have their sides in the simple whole-number proportions described above. The next set of whole-number dimensions which will produce a right-angled triangle is 5, 12, and 13, and the next, 7, 24, and 25. After that, we do not come to another set until 33, 56, and 65.

The question is, is there a rule governing the ratios of the sides of *any* right-angled triangle, and if so, what is it? The first people to have asked this question seem to have been the Greek geometricians, sometime around 500 BC. Exactly who found the answer is not certain; it is always associated with the name of Pythagoras, but it was probably known before his time. Possibly he was the first to *prove* the relationship—of which more later.

In stating the answer to this problem, I shall use the terminology of modern mathematics, not that of the Greeks. The longest side of a right-angled triangle is termed the hypotenuse, and is conventionally denoted by the letter h. The other sides are generally labelled o and a; we shall take o to be the shortest side, ignoring for the moment the fact that there is a right-angled triangle in which the sides enclosing the right-angle are equal. Here we have an elementary example of the use of mathematical symbolism; these convenient labels enable us to talk about the problem without necessarily having to draw diagrams all the time.

The first set of numbers given was 3, 4, and 5. Now, if three is multiplied by itself, the result is nine. This operation is called 'squaring' the number; it is written as 3^2, pronounced 'three squared'. Likewise, $4^2 = 16$. Nine and sixteen added together give twenty-five, and this number, we notice, is 5^2. Very neat, but is it just a concidence? Try the next set of numbers. $5^2 = 25$; $12^2 = 144$; $25 + 144 = 169 = 13^2$. It begins to look as though the result is not coincidence, but the reader is invited to try the same operations on the other sets of numbers given. They will be found to conform to the same general pattern. If this were a new discovery, we should probably want to measure a great many triangles and calculate the results for each, before being certain, but since we are 'cheating' and know the answer already, we can at this stage 'generalise' the results, using the notation described above, and say that in any right-angled triangle,

$$h^2 = o^2 + a^2$$

which, under the name of 'Pythagoras' Theorem' has been drummed into generations of schoolchildren, generally in the form of its plain-language equivalent, 'In any right-angled triangle the square on the hypotenuse is equal to the sum of the squares on the other two sides'.

Having got so far, we can now proceed to forget all about the processes by which we arrived, and remember only the *formula* given above. Then, whenever it is desired to draw a right-angled triangle of any size or proportions whatever, it is only necessary to substitute the figures applicable to the particular case for the letters in the formula, and perform the appropriate mathematical operations.

The result now obtained is good enough for practical men, like builders and engineers. There is a definite rule, and it works. The mathematician, however, is not content with this. He will not be satisfied until he can *prove* the rule, that is, demonstrate by logical and incontrovertible arguments how it is related to the existing body of mathematical knowledge, and why a given procedure not only *does* give a certain result, but *must* give that result and no other. When this has been satisfactorily accomplished, the mathematical statement is dignified by the name of a *Theorem*. Over a hundred different proofs of Pythagoras' Theorem are known, and most of the basic mathematical propositions used in science and technology have been rigorously proved in this way, but there are some which have not been, though they are none the less useful for practical purposes.

This is essentially a book about practical problems, and where it is desirable to introduce a simple mathematical formula, this will be done without considering whether or not it has been formally proved. It must not be supposed, however, that because this kind of approach is possible, the work of the 'pure' mathematician is 'useless'. Far from it. To consider the matter at the lowest level, even the most practical of engineers is much happier working with a formula which he knows has been proved—he cannot otherwise be sure that there is no exception to the rule, however unlikely it may be in practice. Further, in seeking to demonstrate the truth of some proposition, the mathematician must analyse it closely, and will often discover facts and relationships which might otherwise have been overlooked—these discoveries frequently form the basis for further progress. All knowledge is potentially useful, and it is simply not possible to know if, or when, some fact or idea may turn out to have a practical application. Finally, as we know from other walks of life, it is no bad thing occasionally to examine and question fundamental beliefs and assumptions from time to time. By the nature of his work, the pure mathematician is constantly doing this.

MATHEMATICAL SYMBOLISM

Before going on, a brief review of mathematical symbolism and terminology may be useful for those whose knowledge is slightly rusty. We saw above how, when a result is generalised, the numbers in the original problem are replaced by letters, to give a formula which will hold good in all circumstances, irrespective of the actual numbers involved in a given case. Letters are not the only symbols used for this purpose, but generally, for simple formulae, the letters of the Latin alphabet are used as far as possible. In advanced work, this is not nearly enough, and the Greek, German, Hebrew, and even the Russian alphabets have been pressed into service, and the printer's resources of letter design have also been exploited to widen the range. Even so, the number of ideas which it may be desired to express is so vast that it is impossible to assign a single unalterable symbol to each. In general, the meaning of the symbols varies with the problem being studied, and must be defined each time.

There are some important exceptions to this rule. In all mathematical work, certain quantities, ratios, and relationships constantly turn up, which are always the same, irrespective of the nature of the problem. These have been assigned definite symbols, which always bear the same meaning. Examples of two of these *constants* have already been given on page 29—π and e, which are respectively, 3·1416 and 2·7183, to four places of decimals. Other such symbols are ∞, meaning 'infinity' and $\hbar$, the 'rationalised Planck's constant', which occurs in nuclear physics, but which need not detain us here.

Aside from these constants, in principle any symbol can be used to represent any quantity, but to avoid complete anarchy, it has become the accepted convention that frequently recurring quantities shall, within the context of a given subject, be represented by the same symbols always. This saves both readers and writers a good deal of time and trouble, always assuming both to be reasonably familiar with the subject under discussion. So in most branches of physics, temperature is written T, time is t, and mass is m; however, if we are talking about electromagnetism, m represents magnetic moment, and in chemistry it is 'molality'. (It is unimportant for the purposes of the present explanation whether the reader understands these terms or not; let him not be discouraged by them—every specialised discipline has its own vocabulary, which seems incomprehensible at first sight, and the vocabulary of science is no more difficult, when one is familiar with it, than that of, say, cricket or baseball or any other sport. In fact most foreigners find the latter far more difficult.) It is also conventional to reserve letters at the end of the alphabet (x, y, z) for 'unknown' quantities, that is, for those quantities which the formula is to be used to find.

The symbols which have been described so far might be considered as 'nouns' to use a grammatical analogy (or perhaps as pronouns, if the numbers themselves are regarded as nouns). They may be modified ('described') in various ways, as by the use of subscripts, n_3, n_a, or primes, n', N''. Thus, if several different lengths occur in a problem, and are to be incorporated into a formula, it might be decided to call length in general l, and to distinguish the individual lengths as l_1, l_2, l_3, and so on. Provided it is not carried to excess, this is convenient and saves symbols. There are also mathematical 'verbs', which are termed *operators*. Some of them are familiar to everyone: $+$, $-$, $\times$. They instruct us as to what is to be done with the various quantities. In algebra, the $\times$ sign is seldom used; multiplication is indicated by placing the symbols close together, so *ab* is exactly the same as $a \times b$. Division is usually shown by writing the quantities as a fraction; '*a* divided by *b*' is

$$\frac{a}{b}$$

or, if it appears in the body of printed text, *a/b*, which simplifies printing.

Other operators are perhaps less familiar; for example, *n*! means 'multiply together all the numbers up to and including *n*', or, in actual numbers, $6! = 1 \times 2 \times 3 \times 4 \times 5 \times 6 = 720$. Others denote results or relationships: $=$ is familiar to everybody; $\approx$ is less so. It means 'equals approximately' and is doubtless anathema to certain school-teachers, book-keepers, and other like-minded persons with a naïve belief that to every question there is only one possible 'right' answer, but it is frequently found in mathematical expressions which deal with the real world. Similar signs which will often be met with are $>$ 'greater than'; $<$ 'smaller than'; and $\propto$ 'is proportional to'. A group of symbols and operators (termed an *expression*) is often seen enclosed within brackets; this means that an operator outside the brackets acts equally upon everything inside them.

Thus, $a(b + c)$ means 'add *b* and *c*, and multiply the result by *a*'; algebraically the operation is performed directly to give the answer $ab + ac$. Without the brackets, the normal convention 'multiply and divide before add and subtract' would apply, and the result would be $ab + c$. If the reader has any difficulty in seeing this, he should, before going on, substitute figures for the letters and work the example out arithmetically. There may be more than one set of brackets, and the normal 'priority' is $[\{(\,)\}]$.

A very common operation is that of raising a number to a given power, that is, multiplying it by itself a given number of times. One example of this has already been given on page 31 where it was mentioned that a number, say *a*, multiplied by itself once is a^2. A further multiplication by *a* results in a^3, pronounced '*a* cubed', and still further ones in a^4, a^5, etc ('*a* to the fourth, fifth, etc, power'). a^x is '*a* to the power of *x*'. Powers need not necessarily be whole numbers: $a^{2\cdot65}$ is quite a permissible expression, though it is not immediately obvious how one can multiply a number by itself 2·65 times. In fact, this cannot be done by the ordinary arithmetical calculating rules (algorithms), but there are ways of doing it. The quantity e, mentioned above, is frequently found raised to some power, and the power is often given by some quite complicated expression, for instance, $e^{t/(a+b)}$. In modern books this will usually be found printed as $\exp\{t/(a + b)\}$; it is done thus to simplify printing.

This operation is sometimes called involution; the opposite is evolution, or extraction of the root, and is shown thus: $\sqrt{a}$ (the 'square root of *a*'). This is equivalent to the instruction 'find that number which when multiplied by itself gives *a*; to take a simple example, $\sqrt{16} = 4$. Similarly, we have $\sqrt[3]{a}$ ('cube root'), $\sqrt[4]{a}$ ('fourth root'), and so on.

For the actual operations involved in calculating these quantities, the reader must be referred to an appropriate textbook; they have been introduced at this point partly to show the meaning of some symbolism which is frequently encountered, and partly to serve as an introduction to a very useful mathematical convention, which will be used in the remainder of this book.

The ordinary numbers with which everyone is familiar were originally developed for quite everyday purposes: for trade, for counting flocks and herds, for surveying, and for building, and for these purposes they are well adapted. The arabic numerals and the positional system, as extended by the decimal system, serve very well for numbers and sizes between a million and a millionth (1,000,000 and 0·000001), which is good enough for most people, but beyond this they tend to get a trifle unwieldy. Many of the numbers encountered in science and in some branches of technology are much larger or smaller than this, so for a good deal of scientific literature a system is used which saves time and space, and gives a clearer idea of the magnitudes involved.

This system is based upon powers of ten, just like the ordinary positional system, but an indical notation, similar to that described for powers and roots, is used. First of all, it is necessary to realise that when we are dealing with very large numbers, only a few digits are normally 'significant'. If your total savings amount to £1,000 ($2,400), then the loss or gain of £50 ($120) is quite important, but a millionaire is content to know the extent of his fortune within a thousand or two. Likewise, if we are dealing in English billions or American trillions, then an error of a million will still only represent one part in a million, and few instruments can measure that closely.

So, when we encounter a figure of say, six millions, we remember that this equals $6 \times 1,000,000$, which is $6 \times 10 \times 10 \times 10 \times 10 \times 10 \times 10$, and using the notation introduced earlier, we write, instead of 6,000,000, 6×10^6. If the number is, say, 6,290,000, keep to the same principle: consider the first figure as though it were the units, and write $6 \cdot 29 \times 10^6$. In this way it is possible to introduce as many 'significant' digits as desired, but it is seldom that more than three are needed in practice. Since we are dealing in powers of ten, an increase or decrease of one in the 'index' or 'exponent' figure represents multiplication or division by ten. Once one has become accustomed to the method, this gives a very much clearer idea of relative size than do strings of zeros. In this connection it is customary to speak of 'orders of magnitude', thus $5 \cdot 55 \times 10^{10}$ is two orders of magnitude larger than (ie, about a hundred times as great as) $6 \cdot 46 \times 10^8$. Many people (even, it must be admitted, some scientists who should certainly know better) say 'of the order of' when they mean 'approximately'; a loose use of language which is at best unnecessary and at worst misleading.

Notice another useful property of this system: to multiply powers of ten we simply add the indices; thus $10^4 \times 10^5 = 10^9$. If you do not believe this, work it out with the full numbers; you will need no further demonstration of the usefulness of the method.

It remains to extend the system to numbers smaller than one. To do this we make use of another mathematical convention. The *reciprocal* of any number is the result of dividing unity (one) by that number. So the reciprocal of 3 is $\frac{1}{3}$, the reciprocal of 4 is $\frac{1}{4}$, and so on. The convention when dealing with powers of ten is that a minus sign in the index means a reciprocal. Thus, if $10^2 = 100$, $10^{-2} = \frac{1}{100}$ or $0 \cdot 01$, and so on. So, if we have some small number like $0 \cdot 0000026$, we write $2 \cdot 6 \times 10^{-6}$, that is, $2 \cdot 6 \times \frac{1}{1000000}$. Again, when written out in full, the significant digits are remote from the decimal point, which is where they are wanted. The properties of positive indices described above are equally applicable to negative ones, always bearing in mind that, for instance, minus 4 minus 4 is —8, and —5 plus 3 is —2.

FORMULAE IN USE

So far, mathematical formulae have been treated as though they were always, in principle at least, the result of mathematical operations. Pythagoras' Theorem is an example of a formula with many practical applications which can be proved mathematically, and it has been mentioned that there are others which have not yet been proved, but which probably will be one day. However, many formulae used for scientific and technical purposes do not fall into these categories, and some examples will both illustrate this point and serve to show how mathematics is used in the service of science.

Very often, the nature of a formula is inherent in the definition of the measurement or property which we are using it to discover. Such a formula cannot be 'proved' by mathematical reasoning; it is really a shorthand statement which is valid because we have agreed upon it beforehand. Thus, to calculate the work being done in the cylinder of a steam engine, it is necessary to measure the 'mean effective pressure' from an 'indicator diagram' (of which more later), to count the number of strokes the engine makes in a minute, and to know, or to ascertain, the leading dimensions of the engine. Then the following formula gives the answer:

$$H = \frac{PLAN}{33000}$$

where H stands for the 'indicated horsepower', P is the mean effective pressure in pounds per square inch, L is the length of the piston stroke, in feet, A is the area of the piston in square inches, and N represents the number of strokes per minute. Thus a steam engine of 18in bore and 24in stroke, running at 250 revolutions per minute (rev/min) will, if the mean effective pressure is found to be 40lb/in², develop an indicated horsepower of

$$\frac{40 \times 2 \times 254 \times 500}{33000} = 308\text{hp}$$

37

Two points must be brought out in connection with this calculation. The first is that if the engine is a double-acting one, as most are, the number of working *strokes* is twice the number of *revolutions* per minute. Second, the value of 254in² for the area of the piston has been obtained from the figure given for the bore of the engine. This was done by the application of another simple formula, which is familiar to most people. The area of a circle can be calculated, if its diameter is known, by using the formula πr^2 (where r is the radius) or $\frac{1}{4}\pi d^2$ where d is the diameter.

For a second example, consider a completely *empirical* formula. The horsepower which can be safely transmitted by a steel shaft is given by:

$$0 \cdot 02 n d^3$$

Here, n is the number of revolutions per minute and d is the diameter of the shaft in inches. A 3in steel shaft running at 110 revolutions per minute can therefore transmit 59·4 horsepower. This formula could not possibly have been deduced by any process of mathematical reasoning from known facts; it is simply a shorthand statement of the results of numerous experiments. A very great number of the formulae most useful for practical purposes fall into this category, and usually no great mathematical ability is needed to *evaluate* such a formula, that is, to apply it to a specific set of circumstances. The reader should not, therefore, take fright because a writer on technical subjects chooses to present information in this convenient and compact form, instead of using the clumsy and long-winded expressions of 'plain English'. A useful tip, if one finds difficulty in grasping the meaning or significance of some particular formula, is to substitute some very small and simple numbers for the symbols, and work out the result arithmetically. A few experiments of this kind, with different numbers, will soon show how the various quantities of the formula are related to one another, and what effect increasing or decreasing one or other of them has upon the final answer. Once one can use a slide rule (of which more later) there is no need even to stick to small and simple numbers; realistic values can be used, with a consequent gain in understanding.

Adding up the Answer

GRAPHS

Even the least mathematically minded person is nowadays familiar with the general appearance and at least some of the uses of graphs. The reason is not far to seek: graphs offer one of the most convenient and effective ways of presenting numerical information in a compact and easily grasped form. It is assumed that most readers are acquainted with this use of graphs as 'visual aids'; examples readily spring to mind. Temperature charts, graphs of production figures, the state of the country's balance of payments, the rise and fall of the birth-rate, traffic volumes, and the like are encountered every day in business or in the pages of our newspapers.

This use of graphs is by no means unimportant in scientific and technical applications, but here we shall look at some of the less well-known, but very important uses and properties of graphs. First of all, a word or two about nomenclature. In an ordinary graph, of the kind mentioned above, the figures represented along the horizontal axis are called the *abscissae*, and it is conventional that the fixed quantities in a particular case—dates, times, known distances along a route, etc—are plotted along this axis. In mathematical and scientific terms, these quantities are called the *independent variables*—we can, in the context of a particular experiment or set of observations, select them to suit ourselves. Along the vertical axis are plotted the values of whatever it is we are trying to measure or demonstrate—temperatures, rainfall, units of production, etc—the *dependent variables*. The figures plotted along this axis are the *ordinates*. Alternative terms for the axes of abscissae and ordinates are the x and y axes, and these are the terms which will be most used in what follows.

These points are illustrated in Fig 1, which shows something else; namely that it is possible to specify the position of any given point on a plane surface by giving the values of its x and y *co-ordinates* on previously agreed scales. It is quite possible, and for some purposes necessary, to show minus quantities on a graph. As indicated in Fig 1, they are plotted to the left of, and below, the point marked 'O' (for origin). Most graphs encountered in everyday practice, however, will be found to consist only of the 'positive' section above and to the right of the origin. It is also possible to set up a third axis, the z axis, at right-angles to the other two. (Either a model, or some kind of

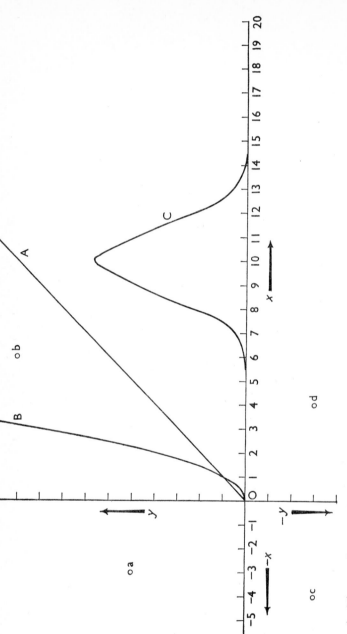

Fig 1 The x and y axes of the Cartesian framework, with various points and lines plotted. The position of point a is $-3x$, $5y$; of b, $6x$, $10y$; of c, $-4x$, $-3y$; and of d, $4x$, $-3y$. Line A is the graph of the equation $y = x$; curve B, the graph of the equation $y = x^2$. Curve C is the Curve of Normal Distribution

perspective drawing, is needed to show this.) It is then possible to define the position of *any* point in space by reference to the three axes. This is the basis of the branch of mathematical study known as Cartesian co-ordinates.

Suppose now the x axis is divided to some convenient scale, as shown. The y axis has purposely been left undivided, so that, for the purpose of illustration, we can make the scale divisions whatever we wish. Let us first make them equal to the divisions of the x axis, and plot a few points at which $y = x$, that is, at, say, the second division along the x axis, we count two divisions up the y axis, and mark a point at the intersection, and so on for other values. If now these points are joined, the result is a straight line, the line A in Fig 1, and at any point on this line, $y = x$. The line is said to be the graph of the equation $y = x$, or, putting it the other way round, $y = x$ is the equation or *law* of the straight line A. If we count two divisions up the y axis for each one along the x axis, we get another straight line, sloping at a different angle. This is the graph of the equation $y = 2x$.

Skipping several stages of argument, it can be said that in general the graph of the equation $y = mx + b$ is a straight line, or, the other way round, that the law of any straight line is of the form $y = mx + b$. m and b are constants which will depend upon the particular line; m determines its slope, and b the point at which it cuts the y axis, known as the *intercept on the y axis*.

The reader would find it interesting and instructive to get some graph paper (obtainable very cheaply from any good stationer), and plot the graphs described, and a few others, for himself. He will in this way learn more about the properties of graphs than can possibly be conveyed in a few paragraphs of a necessarily limited chapter, and being self-acquired, the knowledge is more likely to stick. Let us now take a different rule for plotting the graph: $y = x^2$. To do this properly many more points must be plotted than for the straight-line graphs, but if sufficient are plotted, they can be seen to define a smooth curve, shown as line B in Fig 1. Actually this is only half the curve; if the reader cares to draw the other half, by extending the x axis in the minus direction, as explained above, he will obtain the complete curve. This particular curve is of such importance in mathematics, science, and technology that it is given a name: it is called a *parabola*. Among other things, it represents the path of a projectile (a ball or a shell, say) thrown or fired upwards to land at a distant point.

Finally, as an example of a curve with a much more complicated law, the one marked C in Fig 1 may be instanced. The law of this is

$$y = \frac{1}{\sigma/(2\pi)}\exp{-(x - \bar{x})^2/2\sigma^2}$$

This curve is a tremendously important one for practical purposes, as will emerge shortly; meanwhile we note in passing that σ (a Greek sigma) stands for a quantity called the standard deviation, and $\bar{x}$ means the average value of all the quantities plotted along the x axis (a bar over a symbol nearly always bears this meaning), and, as we should expect, it comes midway along the x axis. This value of x also corresponds to the highest point of the curve.

A great many curves, of differing shapes, can thus be reduced to symbolic form. What is the practical value of this? To answer this, let us consider the following set of figures, which records the results of an experiment in which the temperature of a pan of water was taken at regular intervals as it cooled from boiling point.

Time (min)	Temperature (°C)	Time (min)	Temperature (°C)
0	100	55	46
2	94	60	44·5
4	88·5	65	43
6	84·5	70	42
8	81	75	40·5
10	78	80	39
12	75	85	38
14	72·5	90	37
16	70	95	36
18	68	100	35·5
20	66	105	34·5
25	62	110	34
30	58·5	115	33
35	55	120	32·5
40	52·5	125	32
45	50	130	31·5
50	48		

Adding up the Answer

At first sight, there does not seem to be much rhyme or reason about these figures. Furthermore, if the reader cares to try the experiment for himself, he will almost certainly obtain a different set of figures. We could go on collecting figures like this for quite a long time, without being able to make much use of them. However, suppose we plot them on a graph, which is shown in Fig 2. It is now obvious that there is a definite pattern about the figures, and this general pattern will be repeated if the figures from any experiment of

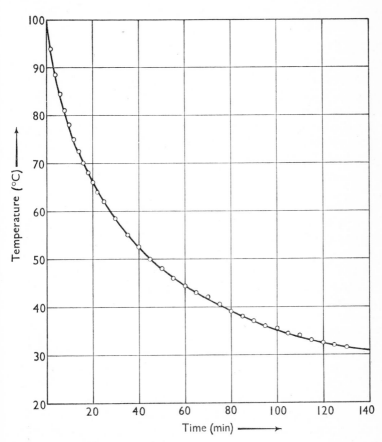

Fig 2 The cooling curve. The points plotted are those of the table on page 42

the kind described are plotted. We may legitimately conclude that, had the experiment been carried out under perfect, or 'ideal' conditions (which in practice, of course, is impossible) all the points would lie on the same smooth curve, which is shown in the diagram.

Now this curve is another one which is well known to mathematicians: its law is

$$y = a^{-cx} + b$$

where *a*, *b*, and *c* are constants depending upon the circumstances of the particular experiment. We can make certain substitutions in this equation, and say that the cooling of the water in the experiment is described by the equation

$$T = a^{-ct} + T_0$$

Here, *T* represents temperature, *t* is time, and T_0 stands for the temperature of the surrounding air. The remaining letters *a* and *c*, are numbers which depend upon the exact experimental set-up. Once they have been determined, by making a few trials, and so long as the set-up is not altered, it is no longer necessary to make measurements or draw graphs; we can predict the temperature at any given time simply by inserting the appropriate numerical values and working out the resulting sum.

This will no doubt strike the reader as an excellent labour-saving device, and so it is, but there is much more to it than that. Because the universe behaves in this way, it is often possible to predict the results of a certain action in advance, and to decide whether or not that action is an appropriate one to take. Thus it is that an engineer in England or Italy can design a bridge to be built in Australia in full confidence that, provided the data given to him were accurate, it will be adequate for the duty required of it. Note, however, that this does *not* do away with the necessity for experiment. The laws must be established by observation in the first place, and there are in an equation generally one or two *parameters* (see page 95) which must be ascertained from actual measurements.

The contribution of the pure mathematician is very well brought out by this example; it is only because he has studied the properties of all kinds of curves, and their laws, that it is possible to look at a graph and recognise that it is governed by a certain equation. There is another reason why it is important that a curve can be described by an equation: for many scientific and technical purposes it is necessary to know either the slope of a curve at a particular point (for example, this represents speed at a given instant in a curve of distance plotted against time), or the area enclosed by a curve. If an equation can be found for the curve, these problems can be solved by applying the calculus, a branch of mathematics rather beyond the scope of this book, but which makes it possible to obtain quick and precise answers where otherwise we should have to be content with laborious and approximate ones.

STATISTICS

Most people have heard of the three grades of lies: lies, damned lies, and statistics, and indeed, no branch of mathematics has suffered greater calumnies than this one. It must be stressed at the outset that this is quite unwarranted. The bad reputation of statistics is due, not to any shortcomings in the subject itself, but to its misuse by people (politicians and advertisers have much to answer for) who either did not understand it themselves, or who understood it well enough but hoped that their audiences would not. The real scientist or engineer regards statistics as he regards the other branches of mathematics, or indeed, any of the techniques described in this book, as a tool. In everyday life, if we are presented with a shoddy job, we blame the workman, not the tool; it is for him to see that his tools are the right ones and to use them properly. Let the reader carry the analogy to its conclusion.

Statistics is extremely important in many branches of science and technology, but it is a vast subject, and the reader who wishes to gain a real insight into it must be referred to the works in the Reading List at the end of this book. One aspect of statistics, the presentation of information by graphical methods, and the deductions which may be drawn in this way, has been briefly treated in the previous section. Here, there is only space to touch upon one other very misunderstood aspect of the matter—probability and the laws of chance.

Many people seem to find difficulty in grasping the idea of probability, which is strange, for it deals with the real world, not with some abstraction of the textbook and the classroom. We all of us know quite well that in real life the answer to a distressingly large number of questions is not a straightforward 'yes' or 'no', but 'maybe'. However, as pointed out earlier, science is quantitative; the laws of probability enable us to assess the degree of reliance to be placed upon that 'maybe'.

It is easy enough to see that if you toss a coin, then, neglecting the fact that one side is very slightly heavier than the other, the chances of its coming down 'heads' or 'tails' are equal. If you toss it a hundred times, you will expect, intuitively, to get fifty heads and fifty tails. When this experiment is tried, it is quite often found that the actual figures are something like 49:51 or 48:52, nearly, but not quite 50:50. However, if a thousand trials were made, the result would probably be 504:496 or some pair of figures close to this, which is much nearer to equality, and the more trials that are made, the closer the figures approach to equality. All matters of this kind, where something is expected 'in the long run' or over a sufficiently large number of samples, fall within the province of statistics.

In the case of the coin-tossing experiment, this result is summed up by saying that the probability of getting a head or a tail in any single toss is $\frac{1}{2}$. Since this represents equal chances, it is easy to see that there is a scale of probability, on which 1 represents absolute certainty; and 0 absolute impossibility. An important practical exercise in statistics is to ascertain the exact probability, on this scale, of any given event. To get an idea of how this is done, let us consider the coin again.

What is the probability of getting heads in two consecutive throws? For purposes of calculation, it is justifiable to treat these as *simultaneous* events; they are consecutive in practice merely as a matter of convenience. Then, since the probability of a 'head' in one throw is $\frac{1}{2}$, the probability of 'heads' in two throws is $\frac{1}{2} \times \frac{1}{2} = \frac{1}{4}$. This is the *law of multiplication*: it is familiar to anyone interested in horseracing, where the odds for an 'accumulator' bet on more than one race are multiplied in just this way. There is also a law of addition, which is used to calculate the probability of an event occurring in one of several possible ways. Thus, with the tossed coin, the probability of getting a head *or* a tail is $\frac{1}{2} + \frac{1}{2} = 1$, which is absolute certainty. Clearly, this must be so, for there are only two possibilities, and we must get one or the other.

Adding up the Answer

To take another simple instance, the probability of drawing any given card from a pack of playing cards is $\frac{1}{52}$. The probability of drawing, say, an ace, irrespective of suit, is

$$\frac{1}{52} + \frac{1}{52} + \frac{1}{52} + \frac{1}{52} = \frac{4}{52} = \frac{1}{13}$$

The two laws may be combined. What is the probability of drawing an ace in each of two consecutive attempts, the first card being returned to the pack before the second is drawn? Obviously, $\frac{1}{13} \times \frac{1}{13} = \frac{1}{169}$. By using the same kind of reasoning, it is possible to predict the probability of a great many events, provided only that all the factors influencing them can be isolated and their individual probabilities assessed.

Once these facts have been properly grasped, it is but a short step to the concepts of combinations and permutations, which play such an important role in statistics, and which will be found fully treated in any elementary textbook on the subject (see the Reading List). Briefly, any problem involving the selection of a group of things from within a larger group, the sizes of the larger and smaller groups being given, is a problem in combinations if the order of the things in the group selected is immaterial, and is a problem in permutations if the order is important. Thus, what is usually called a 'permutation' entry in a football pool is really a combination, since the problem is to select a given number of matches from a list, and their order is not important provided the result is correctly predicted. On the other hand, calculating the number of combinations of letters or figures possible with a so-called 'combination' lock is a problem in permutations, since order is of the essence. Unfortunately there is not space in this book to do more than explain the meanings of these frequently encountered terms; the reader who is interested in either underlying theory or practical applications must be left to pursue his inquiries elsewhere.

A further concept which is related to the ideas just discussed, and which is of great importance in statistics, is that of the average, or *mean*. There are several kinds of mean, but usually, when the term is used without qualification, the *arithmetic mean* is meant. The idea is familiar to most people. Suppose a machinist, on ten consecutive working days, produces the following numbers of articles:

47

Day	1	2	3	4	5	6	7	8	9	10
Output	20	25	50	30	28	31	10	32	26	25

We obtain the arithmetic mean of these outputs by adding them together and dividing by the number of working days. This operation is represented in mathematical symbols thus:

$$\bar{x} = \frac{\Sigma x}{n}$$

The symbol Σ (a Greek capital sigma) means 'add together all those individual values represented by the symbol following'—in this case, the outputs, denoted by x, while a bar over a symbol is the conventional indication that it represents a mean value.

Reverting to actual figures, we have:

$$\frac{20 + 25 + 50 + 30 + 28 + 31 + 10 + 32 + 26 + 25}{10} = \frac{277}{10} = 27 \cdot 7$$

This average, as most of us know, is a very good indication of the ability of a workman. One of its great virtues is that it enables us to discount the occasional very high or very low output, which might be due to fortuitous circumstances, such as a particularly good lot of material or a machine fault, and to view the man's performance with a synoptic eye. This procedure is valuable in many circumstances: it allows us to see the wood in spite of the trees, and to get an overall picture of a situation, but it needs to be applied with caution. It is not suitable for every occasion.

One instance where it is undoubtedly of practical value is in dealing with experimental error. In technical procedures, and even more in laboratory experimentation, one would never, if it is avoidable, rely upon a single measurement or instrument reading. Unfortunately, neither instruments nor observers are perfect, and a series of observations will generally yield a series of different readings. Thus, four separate weighings of a quantity of material might give the following results:

No of weighing	1	2	3	4
Weight recorded (grams)	15·6	15·5	15·4	15·5

What should we accept as the correct weight? In a case like this, we calculate the arithmetic mean, which is here 15·5 grams, and this is taken to be, not the *correct* value, which is unknown (and which, strictly speaking, cannot be known), but the *most probable value*.

We have arrived back at the idea of probability. How did this come about? If a very large number of measurements is taken, and a graph is plotted, with the various individual values as abscissae, and the number of times which each one occurs as ordinates, the result is a curve generally similar to curve C in Fig 1. It will be remembered that this curve was stated to be of great practical importance. The highest point of the curve, corresponding to that figure which turns up most often (and which we can therefore say is most likely to turn up) coincides with the mean value, and we are therefore justified in accepting the mean of a set of measurements as the most probable correct value.

This particular curve is often called a *Gaussian curve*, after the German mathematician C. F. Gauss (1777–1855), who discovered it while investigating the 'Law of Errors', a very simplified version of which has just been set out. The curve had also been discovered independently by Pierre Simon de Laplace (1749–1827), a Frenchman, and Abraham de Moivre (1667–1754), a French-born English mathematician. It was first published in 1733. Each of these workers was investigating something different—de Moivre, for example, was studying the theory of games of chance—and this fact points to the reason for the tremendous importance attached to this curve in statistical theory. It does, in fact, crop up over and over again, in all sorts of apparently unrelated fields, and another name for it is the *Curve of Normal Distribution*.

Among other things, this curve is found to represent accurately the distribution of intelligence among the population. It shows that the vast majority of people have an intelligence which is 'just about middling', which is, of course, what we should expect from our everyday experience. What is not perhaps quite so obvious from everyday experience, is the way the curve slopes away sharply on either side of the mean value, demonstrating that people of very low or very high intelligence are equally rare individuals, and form a very tiny fraction of the population as a whole—which is an awkward situation for those good people who believe that 'all men are created equal'!

There is no space to go more deeply into the matter, but it should perhaps be pointed out in conclusion that none of the subjects mentioned above is a purely academic exercise. Statistical methods are among the most important tools of scientists and technologists. In particular, nuclear physics, biology, psychology, and many branches of industrial inspection rely heavily upon statistical techniques.

COMPUTATION

Enough should by now have been said to show that mathematics is much more than mere arithmetic, which is the only aspect of the subject which most people encounter. Nevertheless, arithmetic, though but a small branch of mathematics, is an important one for practical purposes. Though we can do much with abstract symbols, sooner or later we usually have to get down to figures. This emphasis on the practical aspect is not accidental. Some people—schoolteachers and accountants are frequent offenders—invest figures and calculation methods with a mystique of their own. Many people have heard stories of unfortunate bank clerks working long after their normal finishing time to account for an errant penny among many thousands of pounds. Spurious accuracy of this kind has no place in the workshop or the laboratory, neither have laborious calculations.

This is not to say that slipshod or inaccurate work is to be condoned for an instant. But there is a difference between accuracy and precision. If we are measuring some fundamental physical constant which is to be published for the benefit of other workers, then no pains may be spared to make the measurements as closely as humanly possible, and any calculations made in connection with the measurement must be done with equal care. But if, say, we are putting up some bookshelves, and our best measuring instrument is an ordinary carpenter's rule, then calculations to three places of decimals are a mere waste of time which could be better spent in doing the job.

Therefore, a scientist or an engineer, faced with the necessity to make a calculation, will use a method which will give him an answer to the degree of accuracy needed for the purpose in hand—and no more—in the shortest time and with the least possible effort. One thing which one should never lose sight of, when studying arithmetic in particular, is that although it may be desirable to work out a great many problems in the classroom, this is only because skill in arithmetic, like any other skill, is the result of constant practice. The practice, as such, has no intrinsic value, and if the student is never going to have any use for such facility in computation, there is no point in his acquiring it—he will be much better occupied in gaining an insight into mathematical fundamentals.

Until quite recently, it was necessary for anyone who contemplated a career in science or technology to acquire this computational expertise, for, apart from a few semi-mechanical aids, the human brain was the only tool available for the work. The situation has changed drastically within the last half-century, and we will now look at the modern aids to computation. Nearly everyone has heard of computers, but comparatively few people have much idea of what they are and what they can do, and regrettably, a good deal of nonsense has been spoken and written about them. Every possible computing device falls into one or other of two classes: analogue and digital, and we will deal with them in turn.

Analogue computers

An analogue computer may be defined as a device in which the numerical variables in a problem are represented by some physical quantity. About the simplest device of this kind is the 'Cuisenaire rods' which are nowadays used in enlightened schools as an aid to teaching arithmetic. They consist of pieces of wood of various lengths, variously coloured. The lengths of the pieces are proportionate to the numbers which they represent, so that, for example, a two-unit rod placed end to end with a three-unit rod is found to be equal in length to a five-unit rod, the colours helping in the identification. There is rather more to it than this, but a good teacher, by encouraging children to 'play' with these rods, can lead them to discover the salient facts of arithmetic for themselves—obviously a much sounder method than the 'this is so because I say so' approach.

A rather more advanced form of analogue computing device is the slide rule, which is the constant companion of everyone who has to do much work with figures. A simple mechanical aid to addition can be made by placing two ordinary rulers together; if the zero point on one ruler is arranged to coincide with say, the three-inch graduation on the other, then the result of adding three to any number represented by a graduation on the first can be read off below the appropriate graduation, on the second rule. This sounds a little confusing in words, but if the reader cares to make the experiment, he will soon grasp the principle. Of course, this is trivial; there is not usually much difficulty in adding numbers, particularly those small enough to be contained within the compass of a ruler of convenient length. Multiplication is a much more difficult operation, and one that is more frequently called for in science and technology; the slide rule makes use of the principle just outlined to carry out multiplication.

It was shown above, when dealing with the indical notation, that adding the indices representing powers of a number was equivalent to multiplying the numbers themselves. This is the underlying principle of *logarithms*. The *logarithm* of a number (usually abbreviated to 'log') is the power to which some agreed number (the *base*) must be raised to obtain the number in question. In principle any number can be used as a base for a system of logarithms, but in practice only two are generally used; they are e (which we have met before; it is equal to approximately 2·7183) and 10. Logarithms to the base e, termed Napierian, or natural logarithms, are easier to calculate in the first place, but logarithms to the base 10 are more convenient for actual use, since 10 is the basis of our system of notation. Tables of both kinds are published, but it is not necessary to go more deeply into the subject here. The point is, that if the logarithms of two numbers are added together, the sum will be the logarithm of that number which would result from multiplying the original numbers—in other words the difficult operation of multiplication is reduced to the easy one of addition. Subtraction replaces division in exactly the same way.

Now, the numbers and graduation on a slide rule are set out at distances proportionate, not to the numbers themselves, but to their logarithms. They are arranged on the fixed body of the rule and on a movable slide, and setting the movable slide in a manner analogous to that of the two rulers described above results, not in addition, but in multiplication. It will be seen that the slide rule corresponds to the definition of an analogue device given above, in that numbers are represented by physical quantities—in this case the distances along the rule. A practical slide rule is generally rather more complex than this simplified description indicates, and it can be used for many calculations besides simple multiplication and division, but it is an excellent example of an analogue device, and it has all the advantages and disadvantages of such devices—of which more in a moment.

Now we turn to a rather more complicated device. On page 45 it was pointed out that an important practical problem in many branches of science and technology was to determine the area enclosed by a curve, and also that, provided the curve could be described by an equation, this problem could be solved by the methods of the calculus.

Unfortunately, many curves cannot be described by an equation; one such is shown in Fig 3. This is an indicator diagram; it is obtained from an engine (in the past, usually a steam engine) by causing the pressure inside the cylinder to move a pen up and down against a card fixed to, and moving with, the piston. The practical value of this is that the area inside the line is proportional to the mean effective pressure in the cylinder, and we saw on page 37 that this quantity is needed in order to calculate the indicated horsepower of the engine. The question is, how to measure the area?

A passable approximation to the answer can be obtained by means

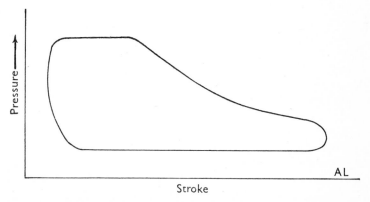

Fig 3   An indicator diagram taken from one end of a steam engine cylinder. The horizontal axis is the piston stroke, the vertical axis represents pressure in the cylinder. AL is the 'atmospheric line', representing the condition of no pressure. In practice, the diagram for the other end of the cylinder is usually superimposed, and is virtually a mirror image of the one shown

of Simpson's Rule, which is used by surveyors to determine the area of an irregularly-shaped plot of land. This entails dividing the area up into strips, which are individually measured; then a calculation according to the rule will give the area. The snag is that to get anything like real accuracy, there must be a great many strips, involving a lot of measurement and a very large calculation, with its attendant drudgery. Therefore, for measuring the area of an indicator diagram, it is usual to use an analogue device, which is called a *planimeter*.

One type of planimeter is shown in Fig 4. To measure any irregular area, it is only necessary to fasten the instrument in a convenient position, and run the moving pointer around the outline. The area is then shown on the dials. There are many practical problems of this kind, which either cannot be defined by a single mathematical formula, or which are only defined by formulae so complex that a solution by orthodox computational methods entails a great deal of work. Provided that all the variables which can affect the answer are known, it is usually possible to build an analogue computer to deal with the problem.

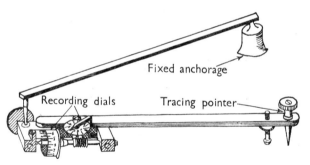

Fig 4   A sketch of one common type of planimeter. Several variations of the basic design exist; for example, the arms may be differently disposed, but the principles, and the basic working parts, of all instruments are similar

Analogue computers are especially valuable in cases where the same calculation has to be carried out many times, using different values for the variables. A good example of a problem of this kind is the prediction of the tides. The factors which influence the tides are well known, but their proportionate effects are different for practically every sea and coastline in the world, and to construct a tide table for only the principal ports of the British Isles involves many thousands of individual complex calculations. One of the first practical analogue computers was designed by Lord Kelvin and built in 1876 specifically to deal with this problem. Various other 'harmonic integrators' and 'differential analysers' were built, up to the mid 1930s. One of the more complex ones was nicknamed 'The Great Brass Brain'. As this suggests, these earlier devices all operated mechanically.

Modern analogue computers are nearly all electrical, the various quantities involved in a problem being represented by voltages, currents, resistances, etc (see Chapter 3 for an explanation of these terms). This has the twofold advantage of a tremendous increase in speed of operation, and a considerable reduction in the size of the machine and the power required to run it. Besides the problems already mentioned, these machines are widely used for engineering calculations, such as the analysis of stresses in structures. In effect, an analogue computer enables us to carry out an unlimited number of experiments, to find the best combination of factors, without having to build an actual mechanism or structure.

The great advantage of an analogue computer is speed. Once the variables of a problem have been set up on the machine, a solution is available as nearly instantaneously as makes no difference. The disadvantages are first, that, as will have been gathered from the foregoing, such computers are essentially one-problem machines, or at best, they can only be set up to solve a limited range of related problems; and second, that solutions are obtainable to only a limited, and predetermined, degree of accuracy. The result of a calculation is generally read off on a dial or series of dials, or some similar device. If the factors affecting a problem should alter, or if the designed degree of accuracy is found to be insufficient, there is usually no alternative to a drastic—and expensive—modification of the machine.

Digital computers

The other type of computer is the digital computer, and it is this that the average person means when he talks about a 'computer'. It might perhaps be remarked in passing that one of the greatest aids to computation is our present system of notation, which denotes

magnitude by position, and includes a sign for zero. We call our notation 'arabic', though in fact it was invented in India, probably in
the first century BC. Who the inventor was is not known, which is a
pity, for this notation must rank as one of the greatest achievements
of the human mind. It is only our familiarity with it which blinds us
to this fact, but it is brought forcibly home to anyone who attempts
to do calculations using roman numerals. In fact, this cannot be done
—the roman numerals, like all the numbering systems which pre-
ceded them, are only a means of *recording* the results of calculations,
which were done, generally, on the *abacus*.

The abacus, or bead frame, was the earliest digital computing
device. Nowadays, in the more highly civilised parts of the world, this
has been relegated to a children's toy, but it still finds considerable
use in more primitive communities. As a matter of fact, a practised
user of an abacus can perform calculations (limited to addition and
subtraction) with considerable facility, and the principle has been
revived, in a more sophisticated form, to make 'pocket adding
machines' which are a boon to those whose arithmetic is a trifle weak
or who, like the present author, are rather lazy.

During the nineteenth century the principle of the abacus was
mechanised, and 'adding machines' (which also subtract) were built.
These were refined and developed into the compact and versatile
desk calculators which find such widespread use in offices today.
These are all digital devices, that is, they handle figures one at a
time (though very rapidly) rather than in the form of continuously
varying quantities, but they are not usually termed 'computers', this
term being reserved for more advanced machines.

The first attempt to build a computer, in the present-day sense of
the word, was made by the English mathematician, Charles Babbage,
between 1812 and 1832. Unhappily, these twenty years of work went
for nothing, for, though Babbage's ideas were along the right lines,
the technology of his time was simply not equal to building such a
machine, which then had, of necessity, to be a purely mechanical
device. Babbage's 'difference engine', still unfinished, is now in the
Science Museum at South Kensington, London.

The next advance was made in 1890 by Herman Hollerith, in the
United States. By this time, electrical technology had advanced to the
point where it was feasible to operate the machine electrically, and

Hollerith's machine was successful. It was originally developed for evaluating census data, and was essentially a sorting and tabulating machine. The central principle of the Hollerith machine, and one which is still widely used today, is the method of recording information on punched cards, which can be 'read' by the machine. These machines proved to be particularly well adapted to the needs of business, and between the two world wars, punched-card machines brought about a revolution in accountancy and office management. By the late 1940s, most large firms had fully mechanised accounting departments.

The true computer was eventually developed, like so much else, under pressure of wartime needs. The correct aiming of bombs from an aircraft presents a very complex problem in applied mathematics, and several ingenious devices were invented to assist in its solution. Finally, the Electronic Numerical Integrator and Calculator (ENIAC for short) was perfected at the University of Pennsylvania in 1946—just too late for actual use in war. As its name implies, this machine worked electronically, and it contained no less than 18,000 thermionic valves (electronic tubes).

This principle of electronic operation was at once realised to be the 'break-through' which had been so long awaited, and progress on both sides of the Atlantic was now extremely rapid—how rapid, may be gauged from the fact that the early machines, built only twenty years ago, are now literally 'museum pieces'; several of them may be seen at the Science Museum. In the 1950s, transistors began to replace valves, enabling faster and more complex, yet more reliable and smaller machines to be built, and development is still proceeding apace, showing no signs of slowing down.

It is out of the question in a book such as this to attempt any explanation of the actual workings of a computer. The most that can be done is to sketch the general principles and a few of the more important uses of the machines. Consider first, then, not a computer, but a human operator setting out to solve some mathematical problem. He will have, either in reference books or in his own memory, the various formulae applicable to his problem, and he will know how to apply them—in other words, he has a 'plan of campaign'. Usually, if the problem is one of any complexity, he will have some kind of mechanical aid, such as a slide rule or a desk calculator, and pencil and paper with which to record intermediate results. Finally, he will be provided with some data—experimental results or something of that kind—which form the basis of the calculation.

We now turn to the electronic analogue of this set-up. A fairly typical computer installation is shown on page 65. It is natural to think first of the electronic equivalent of the mechanical calculating device. In the computer, this is the *arithmetic unit*. Like the desk calculator, it is essentially a device for carrying out basically simple mathematical operations, but freed from the limitations of mechanism, it is very much faster. A single addition in a modern computer takes only a few microseconds, or even less in very advanced machines (a microsecond is one-millionth of a second). To take advantage of speeds like this, the human operator punching the keys of the desk calculator must be replaced by another electronic device, which is called the *control unit*. This is one of the things which distinguish a true computer from a simple calculating machine. In effect, the control unit tells the arithmetic unit what to do. Together they offer the facility for performing a very great number of simple operations, which together may amount to a very complex operation, at very high speeds. The arithmetic unit and the control unit are often regarded as a single entity, called the central processor, and are frequently housed in the same cabinet.

The question of speed of operation must be stressed. A computer cannot do anything that a human being could not do—given time. Unfortunately, human lifetimes are too short for the performance of some complex calculations on a routine basis. Computers, performing in minutes operations which would take a human mathematician months, perhaps years, have made the solution of such problems possible.

So far, we have replaced the desk calculator and its operator by electronic devices. However, in the set-up first described, there was another component, which is, perhaps, easily overlooked, but which is in fact vital. This was the pencil and paper for recording intermediate results. Very few calculations can be carried out in a single operation; more often there is a series of steps. The possession of a means of recording and storing information is the second thing which distinguishes a true computer from a calculating machine. This part of the machine is called the *store*, often referred to colloquially as the 'memory'. Most machines have at least two stores: a slow one, working on the same general principles as a tape recorder, which is used for storing basic information and instructions (of which more presently), and which might perhaps be looked upon as an analogue of the tables and reference books used by a human mathematician; and a fast one, which retains the results and instructions actually being used in calculations. This store must be capable of accepting and reproducing information at a speed comparable with that of the central processor, and much ingenuity has been devoted to the production of faster and faster stores to keep pace with the improved performance of central processors. In general, fast store is the most expensive part of a computer, and accounts for much of the high cost of these machines. This also is the reason why the store is divided; only sufficient fast store to meet the needs of the central processor is normally provided, other information being kept in 'slow' or 'backing' stores, which are much less expensive to build. It should, perhaps, be pointed out that even 'slow' stores are very fast by human standards.

Finally, we must have means of getting data and instructions into the computer, and getting results out. The input and output devices are known collectively as peripherals, and a large machine such as the one shown on page 65 generally has a selection of such devices. Data may be fed in in the form of punched paper tape, punched cards, magnetic tape, or more directly by means of a keyboard, switches, and push-buttons, though these latter methods are far too slow for routine use. Punched cards, magnetic tape, and the rest may be produced by a human operator working a keyboard, but they do not necessarily have to be. Often the data consist of process records, experimental results, and the like, and it is a relatively simple matter to arrange that the measuring and recording instruments shall produce their data directly in a form suitable for computer input, without human intervention. This by no means exhausts the possibilities; already computers can 'read' to a limited extent. If you have an account with one of the large banks, you will have noticed that the numbers of your checks are printed in a curious-looking script which is an obvious modification of the normal arabic figures. They are, in fact, printed in magnetic ink, and a computer can 'read' this script.

Information output may also be in the form of punched cards, magnetic tape, etc, particularly if it is required for the control of other machines, or if it may have to be fed into the computer again at some future time. In addition to this, there will be a printing device, which 'prints out' the answers to problems in normal figures or, in some cases, in a very close approximation to ordinary language.

This completes the description of the major physical parts of a computer, called the 'hardware' in computer jargon. We can feed in the data appropriate for a particular problem, and wait for the answer to come out. We should wait in vain! Something is still missing. Before the machine can perform any useful work it must be provided with the equivalent of the human mathematician's 'plan of campaign', and this the machine cannot do for itself. 'Instructions' have been mentioned several times in the foregoing paragraphs, and the computer's instructions are embodied in a *program* (notice, by the way, that it is conventional, even in Great Britain, to spell this word in the American fashion when it is used in this context).

This necessity has given rise to an entirely new profession: that of computer programmer. It has also given rise to a great deal of misconception. Contrary to what might be expected, the primary qualification of a programmer is not mathematical ability—though mathematical aptitude is undeniably useful. Above all, a programmer must have the ability to think clearly, and to reduce a problem to its essentials. This becomes obvious when it is recalled that the essence of a computer's operations is the extremely rapid performance of a vast number of individually simple operations. This can, perhaps, be made clear by considering a very simple example: the operations of addition and multiplication. The simplest form of arithmetic unit can add numbers and count. If two numbers are fed into it, it will form their sum, which can either be printed out or placed into store, and it can also record the fact that one operation has been completed. Suppose, however, the problem involves the multiplication of say, 356 by 27. It is no use simply feeding these numbers into the computer; left to itself it would add them together to give 383, which is not the required answer.

However, the machine can be given 'instructions' in the form of coded combinations in punched tape, magnetic impulses, or the like, and it can retain these instructions in its store. This constitutes the program for a particular operation. The program for the operation above, translated into plain language would read somewhat as follows: 'Take the first number fed in (356) and place it into one part of the store—call this (a). Take the second number (27) and place it into another part of the store (b). Add the number contained in (a) to the number contained in (a), that is, add the number 356 to itself, and place the answer in a third part of the store (c). Take the number in (b), subtract 1, and replace the answer in (b), cancelling the original number. Then add the number in (a) to the number in (c), and replace the answer in (c), cancelling the original number. Again subtract 1 from the number in (b) and replace the answer in (b). If the number in (b) is greater than zero, repeat the last two operations. If the number in (b) is zero, print out the number in (c).'

If the reader has followed this closely, he will see that it amounts to an instruction to add the number to itself 27 times, that is, to multiply it by 27. For the sake of exactness, it should perhaps be pointed out that a machine as simple as the one assumed would require a separate set of instructions for subtraction, since it can only add and count, but this has been ignored in the example for the sake of simplicity.

However, the essentials of programming are all here: the analysis of a problem into its simplest components, and the instructions to the machine to perform the resulting simple operations in the correct sequence. Of course, only the great speed of computers makes this approach possible. In practice, the operation of multiplication is called for so frequently that this entire program would most probably be retained more or less permanently in the store, being called into operation as needed by the simple instruction 'multiply'. The program is then termed a *routine*. Most modern computers have more complex arithmetic units, capable of performing the operation of multiplication directly; it is then said to be part of the 'hardware'. By contrast, the use of routines such as that sketched is described as a 'software' approach.

Both methods have their advantages. 'Hardware' is expensive, but it is fast, and helps to relieve pressure on the store. 'Software', on the other hand, demands storage space, and means slower operation, but it does offer a means of performing complex operations on a comparatively simple machine. In modern practice, both hardware and software have their parts to play, and it is usual for a computer to be provided, by the makers, with a certain amount of 'basic software', in much the same way as a car comes (or should come) equipped with various standard fittings and tools.

It will be appreciated that a considerable library of software routines may be built up in the course of time, and a machine so provided is a more complex entity (though physically it may be no different) than it was when newly built. It can, to some extent, work by itself, so the description 'electronic moron' applied to computers by some writers is not quite correct. A computer equipped with a large amount of software does display some striking analogies with the human brain, though it would probably be as well not to push these analogies too far in our present state of knowledge. On the other hand, the expression 'electronic brain' is equally unfortunate. In the last analysis a computer cannot think for itself—it has to be told what to do. One occasionally hears of spectacular 'failures' or 'mistakes' by computers, but the safeguards against breakdown or malfunction provided in modern machines are so effective that this is very unlikely—the failures were in the human element responsible (in the broadest sense) for giving the machine its instructions. There is a saying among computer experts which aptly sums up the position: 'Garbage in and garbage out'. It is for the human operators to see that the machine is not provided with garbage, and the human brain alone is fitted for this task. This amounts to 'the human use of human beings' of which we hear so much.

As machines become more sophisticated and complex, the problems, oddly enough, become simpler. The programmer of the earlier computers had to be capable of giving instructions in 'machine language'. Without going into details, it may be said that all digital computers operate in the binary system. Our system of calculation is based on the number ten simply because we happen to have ten fingers, and early man counted on his fingers. However, there is no particular magic in ten, and it is quite possible to devise a system of notation using any number as a base—one such, the duo-decimal system, which uses a base of twelve, is theoretically superior to the decimal system, and finds some use for special purposes. The binary system is based on the number two, and uses just two figures: 0 and 1. This is most convenient for computers, because a simple electronic device has only two possible states; positive or negative, perhaps, or even more simply 'on' or 'off'. A computer is built up of a very great number of such simple devices.

A full description of the binary system would take up too much space, and can in any case be found elsewhere (see Reading List). It is no longer necessary for a programmer to work in the binary system; one of the first improvements was the provision of software (and later, of hardware), which enabled the computer itself to translate decimal notation to binary and vice versa. This trend has continued to the point where programs can be written, and accepted by the machine, in plain, if slightly pidgin English. The language must be carefully learned and correctly used, however; one word has one meaning, and even a complete synonym (in ordinary terms) will be rejected as meaningless by the computer.

The uses of computers are many and various, and new ones are continually being found. They fall, broadly speaking, into two groups, and machines are generally built to suit the needs either of scientific and technical computations, or of business applications—usually referred to as *data processing*. This is generally better than trying to design a general-purpose computer, though some of these do exist. Scientific calculations demand considerable computational capacity —for example, there will be 'hardware' not only for the simple operations of addition and subtraction, but also for multiplication, division, involution, evolution, and possibly others—but they do not, in general, require great storage capacity.

The calculations involved in data processing, on the other hand, are of the simplest description, not usually amounting to much more than addition and subtraction, but large stores are needed. This can be appreciated if it is realised that in order to make the fullest use of a computer in business, the entire records of the business—which would be a large one, otherwise a computer will not be installed at all—must be converted into a form which the machine can use, and placed into store. It is sometimes a surprise to the uninitiated to learn that all information can be converted into numerical form in this way, and it is even more of a surprise to realise that a computer is capable of carrying out, not only arithmetical operations, but also logical operations. The subject of symbolic logic is much too large a one to be dealt with here; it must suffice to say that, when they are reduced to the bare essentials, logical processes have much in common with mathematical ones.

In this connection, advantage is taken of the binary nature of computer operations. Just as all mathematical procedures can be broken down into very simple steps, so can all logical operations be reduced to a series of answers to the question 'yes or no'. This—the existence or non-existence of one of only two possible states—is the simplest piece of information of which it is possible to conceive. It is termed a 'bit', short for *binary digit*. It follows that it is possible to express the 'information content' of a message numerically, and this is the basis of 'information theory', which has uses far beyond the sphere of computers.

Thus, a computer can be programmed to make a decision on the basis of data fed into it. This is a perfectly simple matter, and only seems marvellous because the majority of us do not take decisions on the basis of logic alone, and we are apt to attribute to the machine something of our own complicated and largely emotional thought processes—a great mistake. On the other hand, the computer can only arrive at its decision by way of logical deductions from its data—if the latter are faulty, so will the decision be. The machine is incapable of the kind of intuitive 'leap' which enables a human thinker to compensate for deficiencies in his data, neither can it recognise, as a rule, that a certain piece of data presented to it is inaccurate. The machine cannot create.

Computers can be, and have been, built and programmed to translate from one language to another, to direct traffic, and to set type. These 'logic' applications make a much more powerful and dramatic appeal to the imagination than the purely mathematical ones, but it is the latter which are of more interest in connection with the subjects treated in this book.

Much is heard nowadays about the 'space program', and in the popular mind, this tends to be linked with bigger and better rockets, which are, indeed, a most important factor. But even the design of a rocket calls for some complex mathematics, while the computations involved in putting a satellite into orbit, or in sending a vehicle to the moon or one of the planets, are so complicated, and so vast, that they would demand months of work from human mathematicians. The point is accentuated when, as not infrequently happens, things

Page 65   Two views of a large computer installation at IBM, Hursley

Page 66 A pneumatic gauging installation for checking crankshafts. The component to be checked is clearly seen in the view on the right, showing the loading position. Once loaded, the component is swung over, as shown on the left, when all diameters can be measured simultaneously, and

do not go quite according to plan. Then, fresh calculations may have to be made in a hurry, and the answers are needed quickly if they are to be of any use—so quickly that only computers can deal with the job. Attention is now being directed to the problems involved in building a computer small enough, and robust enough, to be sent aloft in a rocket.

Another aspect of science and technology which has impinged forcibly upon our everyday lives is nuclear research. Many people may think it better that we should be without it, though on balance, probably more good than ill has come of it. However that may be, this is another field which poses the most formidable mathematical problems, and the use of computers has not merely lightened the burdens of the theoretical workers, it has made it possible to undertake pieces of research which would otherwise have been quite out of the question.

3

How Large is it?

At the beginning of the last chapter, it was pointed out that *quantitative* methods of thinking and expression were fundamental to science and technology. In order to make quantitative statements about anything, two things are required: a method of measuring those aspects of the object in which we are interested, to the degree of accuracy required for the purpose in hand, and some means of recording the measurement. It might conceivably be argued that the means of recording are not an essential feature of the operation; thus, generations of bygone craftsmen successfully made parts which fitted one another, purely by comparison. This is really a question of what we mean by the word 'measurement'. Turning to the *Concise Oxford Dictionary*, we find that it is the noun associated with the verb 'to measure', which in turn is defined as to 'ascertain extent or quantity of (thing) by comparison with fixed unit or with object of known size'. It will be obvious that the need for recording is implied in this definition, which is the one which will be assumed throughout this book, without further philosophical justification.

Before we leave philosophy for practice, however, one other thing ought perhaps to be pointed out, which, though it may seem to be of merely academic interest at the moment, is something which can have the profoundest significance. The very idea of measurement, as defined above, assumes certain things about the nature of the universe in which we live, which, though they may be matters of everyday experience cannot be said to be 'proved' in the strictest sense. Mathematicians can imagine a universe (or more accurately, several universes) in which some or all of these things are no longer so, and there are reasons for believing that the actual universe does not entirely conform to the rules of 'everyday experience'.

How Large is it?

THE ORIGINS OF MEASUREMENT

Men were certainly measuring things long before the date of any of the records which survive today—probably before the invention of writing itself. We can only conjecture about the first measurements, but it is not difficult to see how they might have come into being. The need would have been felt almost as soon as agricultural societies had begun to supersede the earlier hunting and pastoral ones—possibly even earlier.

It is a reasonable supposition that measures of length came first, not only because they would early have been required for the purposes of the market place, but because everybody carries a set of convenient yardsticks about with him, in the various dimensions of his own body. People vary, of course, but the majority of adult persons do not differ greatly in size and shape, and this method of measuring would have been sufficient for the needs of the time. It also happens, rather conveniently, that certain of the basic dimensions of the human body are related to one another in a way that enables us to build up a quite simple and logical set of measurements.

One of the easiest ways to measure something in a rough-and-ready way is to compare it with the distance from one's elbow to the tips of the extended fingers. This is the cubit, which will be familiar to all readers of the Bible. In modern terms it is equal to about 18 inches or 46 centimetres. There can be few of us who have never had occasion to make use of the fact that the pace of the average man is about a yard long. Here is another basic measurement, which, most usefully, is double the cubit and about three times the size of another very widely used standard—the length of the foot. Proceeding, we find that two yards is practically equal to the fathom—the distance from fingertip to fingertip of a man's fully outstretched arms. This dimension is, in a normal man—can it all be coincidence? —equal to his height. For smaller measurements we can turn to the breadth of the hand, which is about four inches, and is still used as a standard for measuring horses, while the lengths of the thumb joint or first finger joint are the undoubted forerunners of the inch.

It is perhaps not surprising that evidence of the use of measurements such as these has been discovered in every ancient culture so far investigated, or that, adapted as they are to human scales and human needs, they should have remained in use, virtually unchanged, from the earliest times until the present day—a period of some seven thousand years.

Measurements of volume, of capacity, would also have been required at an early date for measuring grain, wine, oil, and similar commodities. It appears logical enough to us to define the unit of volume in terms of the unit of length, but there would at first have been severe difficulties in doing this. It is no easy matter to make a cubical container, completely watertight and to the necessary degree of accuracy. The first standard measures of volume were cylindrical pots—the potter's wheel was one of the earliest mechanical inventions—and they must therefore have been purely arbitrary. This implies a fairly high degree of organisation in the communities which produced and used them, for there would have to be either general agreement upon the standard or, more likely, a chieftain or king strong enough to impose it. Thus, the ancient Egyptians had a unit of volume, the apet, based upon the wine jars of the Pharaohs, which was certainly in use 2,000 years BC.

Units of weight have very little meaning without an accurate method of weighing, for, as anyone who has tried it will agree, human judgement is very poor in this respect. Such units cannot therefore have pre-dated the invention of the beam balance. The date of this is uncertain, but good and accurate balances and sets of weights have been found in Egyptian tombs and dated to 1500 BC, so the invention is at least as old as that. It seems likely that the intrinsic accuracy of the weighing process was recognised very early, for precious metals and drugs were among the first commodities to be weighed; indeed, weighing was for a long time used only for comparatively small quantities. The units of weight seem to have been more or less arbitrary, though there is some reason to think that they may have been based upon the weight of a grain of corn. Certainly this would explain the close similarity of the weights in use in various early civilisations.

Time was not nearly so important in earlier cultures as it is—or as we have made it—today. Like so many other things, the calendar was first developed in Egypt. The agriculture of the ancient Egyptians depended upon the annual inundation of the Nile, and it was most

70

desirable to have some means of knowing when it was about to occur. The phenomenon is a very regular one, and in principle this could be done, as no doubt it actually was at first, by counting the days from one inundation to the next, say by carving notches in a piece of wood. But this is a notoriously unreliable method without some independent check, and even before the dawn of history the Egyptians were using the Sothis cycle for this purpose. That is to say, they reckoned their year from the date on which the rising of the sun coincided with that of Sirius, the brightest star in the sky, which they called Sothis. It would not have taken many years of astronomical observations to show them, if indeed they had not realised it already, that the solar year (on which the seasons depend, of course) does not consist of an exact number of days, and they soon adopted the device of intercalary days, which survives, with but slight refinements, in our Leap Years.

The ancient civilisations of Mesopotamia also evolved in flat, relatively cloudless lands, where men's attention would have been early directed to the stars, and they also developed accurate calendars at an early date. To the Babylonians also we owe another fundamental invention, the sundial, and the division of the day into 24 hours stems from this time. Without going into too much technical detail, it can be said that this is purely a matter of geometry: it is an easy matter to divide a circle into twenty-four (or any other even multiple of six) parts by stepping out the radius around the circumference and repeatedly bisecting the resultant angles, simple operations which can be done with ruler and compasses only.

During the time we have been discussing, and for a long time afterwards, the measurement of temperature was purely subjective. We will, however, defer detailed consideration of this until Chapter 5. Let us now see how the earliest standards were gradually developed and refined into those we use today.

ABSOLUTE STANDARDS

The need for more accurate standards of measurement, and the ability to produce them, went hand in hand with the development of technical skill. Kings and governments caused national standards of length, volume, and weight to be constructed and maintained, and the working standards of craftsmen and scientists could (in principle anyway) be compared with these and adjusted. However, for the practical man, and even more for the developing needs of science, this was not really enough. Workers in different countries could never be sure that they were using the same standards of measurement, even where these went under the same names. Even within a country doubts began to be felt as the ability to take measurements became more and more refined. The need for absolute standards was beginning to be felt. Only in the measurement of time was appeal made to the unchanging motion of the stars (or more precisely, of the earth). All other standards were completely arbitrary, and if the master standards were destroyed, there was no possible way of guaranteeing that the new ones would be exactly the same as the old.

This was the state of affairs when the French Revolution ushered in the so-called Age of Reason. The spirit of the times was propitious, and the newly-formed Academy decided to introduce a new system of measurements, which was to be based upon an absolute natural standard—the size of the earth. This in fact was a bad choice, for it cannot be measured by any direct method, and even with modern resources its exact determination is an appallingly difficult business. As a matter of fact, the exact measurement, which is essential for the correct aiming of intercontinental missiles, has only quite recently been made, and is regarded as 'classified' information—quite why, it is difficult to see, since any nation with an interest in the matter has presumably made the measurement.

Be that as it may, the basis of the 'metric system' was stated to be a measurement of length, supposed to be one ten-millionth part of the distance from the North Pole to the Equator, measured on the meridian of Paris. This is called the '*metre*' (in US, *meter*). Larger and smaller units are derived from the principal unit by way of powers of ten, thus making calculation by decimals very easy. It is assumed that the reader is familiar with the principal units of the metric system as it is used today all over Europe and in many other parts of the world, and as it is soon to be introduced into Great Britain and perhaps eventually into the rest of the English-speaking world, and this brief historical sketch will be concluded upon this assumption. The unit of volume was defined in terms of the unit of length and the properties of pure water, again thought to be a natural standard. Thus, one cubic decimetre of distilled water at its temperature of maximum density (4° C) was the litre. Finally, the unit of mass, the gram (or gramme) is the mass of one cubic centimetre of pure water at the same temperature.

All these definitions have grave theoretical shortcomings. We shall not go into these here, but interested readers can pursue their studies after consulting the Reading List. A word of warning, however. For scientific purposes it is very necessary that the theoretical and practical shortcomings of the metric system be known and understood. Unfortunately, some modern writers, in discussing these shortcomings, do tend to give the impression, perhaps unwittingly, that the originators of the metric system rather bungled the job. This is far from the truth, and it is unfair to criticise the origins, as distinct from the form, of the system, in the light of knowledge gained in the subsequent 150 years or so, largely from better methods of measurement than were available in 1799. These improved measurements have shown that the inter-relationship of the various units of length, capacity, and mass is not exact, and for certain scientific purposes this has to be taken into account. But for most scientific, and for all everyday purposes, the system as defined is perfectly adequate.

Nowadays, two self-consistent systems of measurement, both based on the metric system, are in use for scientific purposes. These are the centimetre-gram-second system (cgs) and the metre-kilogram-second system (MKS) [these ways of writing the abbreviations are conventional]. Starting with the units named, a complete system for all purposes can be built up. Of these two, the MKS system is theoretically more sound, and is being adopted more and more. An absolute standard of length, from which the other units can, at least in theory, be derived, has been discovered, and can be explained, briefly, as follows.

It is well known that any substance, if sufficiently heated, will give out light. With many elements (see Chapter 4 for the definition of an element), this light is of a very definite colour, and the reason for this is that the light given out by the heated element has only one or two definite wavelengths. Light is a wave motion; the exact nature of the waves is a question which we shall avoid for the moment, but

73

the point is that these waves are very small, and are always of the same length for a given element. This effect can be demonstrated by sprinkling a little salt into a gas flame; a brilliant yellow colour will be seen. This is due to the element sodium; exactly the same colour will be obtained if some other sodium compound, such as washing soda, is used, and the compound need not even be particularly pure. Provided we can measure the wavelengths, therefore, we have here a simple and invariable standard of length. This measurement can be made by modern techniques, and the length of the metre is defined as being equal to 1553164·13 times the wavelength of a certain red light emitted by the element cadmium. The separation of the various wavelengths, if there is more than one, is done by means of the spectroscope, more fully described in Chapter 4.

The smallest unit provided for in the original metric system is the millimetre, but for certain purposes it has been found convenient to define three smaller units: the micron, generally written as a Greek letter mu (μ), which is 10^{-3}mm; the millimicron (mμ), 10^{-6}mm; and the Ångstrom unit, 10^{-7}mm.* The latter is used for measuring the wavelength of light or other short-wave radiations (mμ are also sometimes used for this), while the sizes of many microscopic objects, such as bacteria, are conveniently measured in microns.

MEASURING METHODS AND DEVICES

Measurements of weight are the easiest to make accurately. Strictly speaking, weight is measured directly only by the spring balance, a somewhat crude, but very convenient instrument. The principle is embodied in many domestic scales and similar appliances, where simplicity and robustness are of greater importance than extreme accuracy. Accurate weighing is done with a beam balance, and this really measures, or rather compares, *masses*, not weights. An explanation of the difference can be found in physics textbooks, but for our purposes we are justified in regarding weight and mass as the same thing. The principle of the beam balance is shown in the sketch, Fig 5. A laboratory balance will have various refinements not shown here, such as a means of relieving the knife-edges of load when not actually in use, a pointer for indicating the oscillations of the beam,

* These are the commonest terms at present, but see the appendix.

and perhaps a device for damping these oscillations and bringing the beam to rest quickly. A sensitive balance is generally enclosed in a glass case so that air currents cannot disturb it.

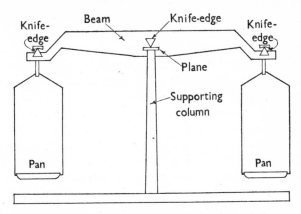

Fig 5   The essential working parts of a beam balance

In use, the object to be weighed is placed in one of the pans. It is customary to use the left-hand pan for the object. Standard weights (usually made of brass, though other metals, possibly even platinum, may be used for better-quality ones) are then systematically added to or removed from the right-hand pan until equal oscillations of the beam show that the two pans are in balance. An ordinary laboratory balance, as used by students, is sensitive to a weight of 5 milligrams (mg). If that does not sound very impressive, remember that the weight of an ordinary aspirin tablet is 300mg. A really good chemical balance, however, is sensitive to 0·1mg, while the special balances developed for 'micro-chemistry' can weigh to 0·001mg.

These instruments, which are so sensitive that the heat of the operator's hand can affect the weighing if due precautions are not taken, represent the ultimate in the development of the kind of balance shown in the diagram, but even their performance pales into insignificance beside that of the latest 'deci-micro' balances. One of these is shown on page 84. The knife-edge bearings of the conventional balance are replaced, in these instruments, by flexure bearings consisting of quartz fibres, and the result is a comparatively robust

75

instrument which is accurate to within 10^{-8}g. It is difficult to provide an illustration which will give the reader any idea of the smallness of this quantity. The weight of the ink in a single full stop on this page is about eighty times as great!

All weighing appliances are basically either beam balances or spring balances, but many modifications have been made from time to time to suit various special purposes. Thus, the steelyard, as illustrated on page 83, and which is particularly associated with the Romans, was invented to meet the demand for a comparatively portable instrument for trading purposes, and without the inconveniences of loose weights. The object to be weighed is placed in the scale pan, and the single weight is slid along the beam until the latter is horizontal. The weight can then be read off on the graduated beam. The instrument depends upon the principle of leverage, of course, and by using a system of interconnected levers, instead of a single one, we can extend the principle to build a 'weighbridge', on which objects weighing many tons can be balanced against a comparatively small standard weight—a great convenience. The principle may be seen on a small scale in the weighing machines found in most chemists' shops.

A different application of lever principles was discovered in 1669 by a Frenchman, Jean de Roberval. For a long time his principle was not understood, and was referred to as 'The Static Enigma'. The principle is sketched in Fig 6. There is no room here to explain the

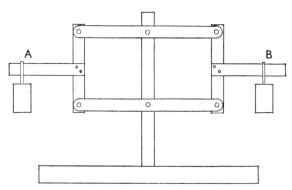

Fig 6   The 'Static Enigma' of Jean de Roberval—the working principle of most tradesmen's scales

principles involved, which can be followed in any textbook of mechanics by those sufficiently interested. The practical point is that, provided the two horizontal arms and the two vertical arms remain parallel to themselves, a condition fairly easy to secure, then equal weights placed on the extensions A and B will balance irrespective of their exact positions along the arms. Using this principle, it is possible to construct a robust, yet compact mechanism more suited to the needs of trade than the comparatively delicate beam balance, and the majority of counter scales are still built on this principle today.

Measurement of volume

The principle of volumetric measurement must be familiar to anyone who has done some cookery (or watched it being done). The appliances and methods used in the laboratory are essentially similar to the kitchen measuring-jug or cup. They are somewhat more accurately made and graduated, but the measurement of volume is not susceptible of the intrinsic accuracy of weighing, with which we have just been dealing. An experienced worker, using good apparatus, can make volumetric measurements accurate to within one part in a thousand, but a certain amount of luck is involved. An error of 0·2 per cent is accepted as normal for routine analyses using volumetric methods. This contrasts sharply with the one part in ten thousand error which an advanced student would be expected to attain with a reasonably good balance, and it is a far cry indeed from the one-in-ten-million accuracy attainable with some of the modern balances discussed earlier.

Thus, when the most accurate results are wanted, even liquids are weighed in the laboratory, but for everyday work, volumetric methods have the advantage that they are quick and convenient. Volumetric measurements are most often made for chemical purposes, and in what follows we shall assume that this is being done, but of course such measurements can be, and are, used for other scientific and technical purposes.

The simplest instrument is the measuring cylinder, which is little more than an accurate version of the measuring cup mentioned above. It is made of glass, and is graduated usually in cubic centimetres (cc), sometimes referred to as millilitres (ml). The cylinders used in laboratories are generally straightsided, but pharmacists, and sometimes photographers, use measures of conical shape, which have the advantage that the graduations at the bottom, where smaller quantities are involved, can be more closely sub-divided. Chemists, however, turn to different instruments when more accurate measurements are required.

Thus, when it is desired to measure an accurately known quantity of liquid, as when making up a standard solution, a measuring flask is employed. This is a flask with a long neck, which holds an exactly stated amount of liquid when it is filled up to a mark engraved on the neck. Figure 7 shows such a flask, and also illustrates a most important point in the taking of volumetric measurements. The surfaces of most liquids contained in glass vessels are not perfectly flat, as might at first be thought, but are slightly concave, due to a kind of surface 'skin', termed the *meniscus*, which clings to the sides of the vessel. The formation of the meniscus is a somewhat complex phenomenon, into which we shall not enter here, but it can quite easily be seen if some water is poured into a fairly narrow vessel, such as a wineglass. All measuring apparatus is graduated so that the true reading is obtained when the bottom of the meniscus is at the mark. Moreover, the mark must be at eye level if an accurate reading is to be obtained. Measuring flasks should only be used at a temperature close to that at which they were originally calibrated, usually 15° C. They are made in various sizes, from 100cc up to 2,000cc (2 litres), the size most used being 250cc.

A measuring flask is made to *contain* a definite volume of liquid, but it will not *deliver* that volume, for there will always be some liquid left clinging to the sides of the flask. To measure out stated volumes, pipettes and burettes are used. A pipette (Fig 8) is intended to deliver a definite stated volume, usually 25cc, though other sizes are made. In use, the liquid is sucked up into the pipette until it is just above the mark. The shape of the pipette ensures that the liquid will not rise too fast, thus enabling some control to be exercised over the operation. Even so, only liquids known to be harmless are drawn up by 'mouth suction'; a safety device such as a rubber bulb or a vacuum bottle should always be used with caustic or poisonous liquids. The pipette having been filled with liquid, the latter can be retained by placing a finger over the top of the pipette. The liquid

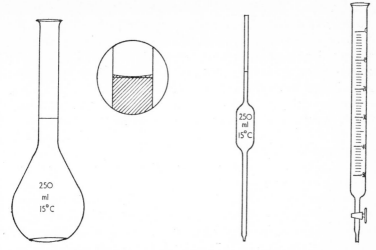

Fig 7　A measuring flask, with (*inset*) the appearance of the liquid
meniscus when the flask is correctly filled up to the mark

Fig 8　A pipette

Fig 9　A burette

level is then allowed to drop to the mark by easing the pressure of
the finger, when the surplus liquid will run out at a controlled rate.
The charge of liquid can then be conveyed to where it is wanted, and
discharged by removing the finger closing the pipette.

When it is desired to deliver definite, but varying quantities, or to
measure a quantity actually delivered for some purpose, as in 'titration', which is described in Chapter 4, a burette is used. This is
illustrated in Fig 9, which is practically self-explanatory. The required quantity of liquid is run out by opening the tap, and the
amount delivered can be read off on the graduations. Burettes are
usually graduated in tenths of a cc, and the majority of them are
arranged with the zero graduations at the top, so that a direct reading
of the amount delivered, rather than of the contents of the instrument,
is obtained.

Before leaving the subject of volumetric measurements, it should
be mentioned that instruments for measuring the volumes of gases
are also made, though lack of space precludes detailed description

of them here. In general they are of two forms; laboratory instruments depend upon the fact that a gas at a definite pressure will displace a definite volume of liquid, which can then be measured directly. For industrial purposes, where robustness and simplicity of operation are important, instruments are employed which embody a pair of metal bellows. These are alternately filled and emptied, their movements being recorded directly on some sort of dial. The domestic gas meter is a good example of this kind of instrument.

Measurement of length

Measurements of length, or of quantities based upon length, such as areas and (by the application of trigonometry) angles, are of great importance, not only for scientific, but also for technical purposes. In fact, the study of measurement, basically of this kind, is practically a science by itself. This study is called metrology. Considerable accuracy can nowadays be attained, and is in fact essential, not only in the laboratory, but in many technical applications as well. For example, the mould used to cast the type for printing this book was built, as a routine manufacturing process, to limits of 0·00001in. The type itself is produced to limits of 0·0001in—one ten-thousandth of an inch. By contrast, the diameter of a human hair is about two thousandths of an inch, so a 'hair's-breadth' in this context is a very coarse measurement indeed!

The most obvious, simplest, and also the least exact way of measuring the length of something is to compare it with something else of known length. This method needs no description or elaboration; even the least technically minded person has frequent occasion to use a ruler or tape-measure. Because of the limitations of the human eye, this is not a very exact method; a reading to within 0·01in, or say 0·25mm, is about the best that can be hoped for even under quite favourable conditions. Indirect comparison gives better results, provided suitable standards are available, and to see how this works, it will be necessary to introduce the important principle of 'end measurement'. Numerous devices have been made for this purpose; a simple slide gauge will be considered here.

Suppose we have a straight bar, with a second, shorter bar fixed to one end of it exactly at right-angles. (The reader may find it easier to follow this description if he refers to Fig 10, which is dealt with in more detail later.) If now a third bar is also arranged at right-angles to the long bar, but in such a way that it can slide along the latter while still remaining at right-angles and thus parallel to the other short bar, it is obvious that we have here a ready means of gauging the length of an object, simply by placing it between the two 'jaws', and sliding up the moving one until the object is just gripped. If then we had a number of bars of different known lengths, we could ascertain the length of the object by finding which of them would fit between the jaws with the same degree of tightness. In this way it would be possible to measure to within 0·001in—it is a fact that in this respect the sense of touch is much more sensitive than that of sight.

However, although this principle is actually used, in a refined form, for comparing measuring instruments with standard measures, it would be far too tedious a process for everyday use. Thus the bar of the slide gauge is graduated directly in inches or millimetres, but this of course brings us back to the limitations of the human eye. Fortunately, there is a way of overcoming the difficulty, and as this involves one of the oldest and most important principles of precision measurement, it is worth taking some trouble to understand it. The principle was introduced in a practical form by Pierre Vernier (1580–1637), and is accordingly known as the vernier principle, though it had been described earlier by the mathematician Christopher Clavius, around the end of the sixteenth century. It depends upon a peculiarity of the eye: although it is difficult to see small differences in *size*, quite tiny differences in *relative position*, or displacement, are fairly easy to detect.

Referring now to Fig 10, we see the fixed jaw and the beam of our slide gauge, which in this case has been divided into tenths of an inch. However, on the moving jaw, at the right, another scale has been placed, in such a manner that it comes close to the main scale. This 'vernier scale' has ten divisions, and its length is such that it covers just nine divisions of the main scale. If the first graduation on the vernier scale coincides with a graduation on the main scale, indicating that the jaws are set an exact number of tenths of an inch apart, then the final graduation on the vernier scale will also correspond with a graduation on the main scale.

Now, since nine divisions on the main scale equal ten divisions on

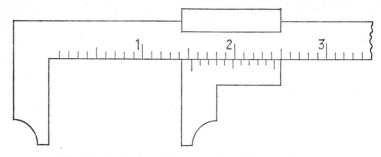

Fig 10   A vernier slide gauge to read to 0·01 in

the vernier, the difference between one division on the main scale and one division on the vernier is one-tenth of one-tenth, or one-hundredth (0·01) of an inch, and the second graduation of the vernier will fail to coincide with the second graduation of the main scale by that amount; the next two graduations will not coincide by 0·02in, and so on down the length of the vernier scale. So the rule for reading the vernier scale is: note the number of complete inches and tenths to the left of the zero mark of the vernier scale, then see which graduation of the vernier coincides with a graduation on the main scale; the number of this is the number of hundredths which must be added to the reading of the main scale. In the illustration the instrument is shown set to read 1·43in, and it will at once be seen how simple it is to observe the coincidence of lines, as compared with trying to read a scale graduated directly in hundredths of an inch—a task practically impossible to the naked eye.

As a matter of fact, an instrument reading to 0·01in has been taken as an example partly to simplify the explanation, but mainly because of the limitations of the printing process, which make it difficult to reproduce very fine graduations. It is a comparatively simple matter to make an instrument capable of readings to 0·001in, and such 'vernier calipers' are in everyday use in engineering workshops.

As has been pointed out above, a measurement of 0·001in cannot be considered to be particularly precise by present-day standards. To attain greater precision, different principles have to be adopted, and the method which finds the most use today is based upon the screw thread. Suppose we take a screw having, say, ten threads to an inch,

Page 83 A Roman steelyard found at Pompeii, and therefore known to have been made prior to AD 79. Compare with the picture on page 84

Page 84 The Oertling model Q01 deci-microbalance. This represents nearly nineteen hundred years of development from the balance shown on page 83. The balance pans are here contained within the fluted columns, which are actually thermally insulated covers. They are raised when access to the pans is necessary, and remain closed at all other times, to protect the pans and the sample from atmospheric or temperature effects. Once the balance has been calibrated, the weight of a sample can be 'read out' directly from the scales seen in the upper part

working in a well-fitting nut, which can be fixed in some convenient position. Now, turning the screw through one complete revolution will result in its moving forward one-tenth of an inch. It is evident that here is a means by which we can very easily make measurements to a high degree of accuracy, for if a disc is mounted on the head of the screw, and its periphery divided into ten parts, we can make tenths of a turn, and thus measure hundredths of an inch. If the disc is one inch in diameter, then the graduations representing 0·01in will be (in round figures) one-third of an inch apart, and the speed and convenience with which readings may be taken will be obvious.

Measuring appliances working upon this principle are a later development than the vernier; they had to wait upon methods of making accurate screw threads. It is striking how, time and again, the advances in technology which make an invention necessary go hand in hand with corresponding advances which make it possible. James Watt (1736–1819) designed an instrument of this kind, but with the crude methods of his day, engineers could not make it; neither would they have had any use for such accuracy. However, Sir Joseph Whitworth (1803–87), who during the mid-nineteenth century played a large part in the development of modern engineering techniques, and did important work in the standardisation of screw threads, built a 'measuring machine', using a screw thread, which was capable of measuring to within one-millionth of an inch. At about the same time, in 1848, the Frenchman, Jean Palmer, invented the micrometer, which is today virtually the 'standard' measuring instrument in engineering workshops.

The basis of a micrometer for inch measurements is a very accurate screw, with forty threads to the inch. This is connected at one end to the actual measuring member, the design of which varies according to the kind of measurements which it is desired to make, and at the other end to a 'thimble', which has 25 circumferential graduations. It is thus possible to make a measurement of one-twenty-fifth of one-fortieth, that is, $0.025 \div 25 = 0.001$in. As a matter of fact, a skilled worker can take a reading of 0·0001in without much difficulty.

Still more accurate measurements can be made by using optical methods. Most of these depend upon the 'interference of light', a phenomenon which is described in textbooks of physics. Typical of devices of this kind are the instruments developed at the National

F

Physical Laboratory, Middlesex, England, and at the National Bureau of Standards, Washington DC. These make use of 'diffraction gratings', which are sheets of glass (or some other transparent substance) ruled with very fine lines, some 500 per millimetre. If two such gratings are superimposed at a slight angle, a pattern of light and dark bands will be seen, and a very tiny movement (of the order of 0·001mm or less) of the gratings relative to each other will be reflected in a comparatively enormous movement of the bands, quite large enough to be easily visible to the naked eye. This principle has been applied in building automatic machine tools; the movement of the light bands is detected by a photo-electric cell, and the electrical output from this, suitably amplified, is used to control the machine slides.

There is one kind of linear measurement which is very important in engineering practice, but which has always presented great difficulties. This is the measurement of circular bores. Very few mechanisms do not incorporate one or more of these, and in some cases (the cylinders of a motor-car engine are a good example) they must be produced and measured to a high degree of accuracy. The work of measuring these was time-consuming and called for considerable skill; moreover, the nature of the measuring method, which consisted basically of introducing a closely-fitting plug into the bore, led to wear of the gauges, small, it is true, but sufficient to be of importance in some cases.

This difficulty has been overcome in an invention developed, like so many others, under the necessities of wartime, but only quite recently brought to its fullest usefulness. This involves an extremely ingenious principle known as 'air gauging' or 'pneumatic gauging'. As applied to the measurement of holes and bores, this consists of a plug, made to an accurately known size, but smaller than the bore by a small amount. This is hollow, but closed at one end, and compressed air is introduced into it by a connection at the outer end. The air can escape through holes in the periphery of the plug. So long as air can escape freely, no pressure will be indicated by a gauge connected to the air supply pipe, but if the escape of air is restricted, the pressure will rise in proportion to the restriction.

To use the device for gauging a bore, it is simply introduced into the bore, and the air pressure is switched on. Since the plug does not exactly fit the bore, a certain amount of leakage will take place, and it is found that first, the amount of leakage (and thus the pressure recorded) is exactly proportional to the clearance between the plug and the bore, and second, that for very small differences in clearance, comparatively large changes of pressure are observed. The size of the plug being known, it is a simple matter to calibrate the pressure gauge directly according to the amount the bore is oversize, and we then have a very efficient and rapid means of measuring the bore. Moreover, since the plug does not actually touch the walls of the bore, there is no wear to worry about. In the latest development, the principle has been applied to the measurement of external, as well as internal dimensions, and on page 66 a specially-built equipment is seen checking all the dimensions of a complicated component in one operation, to a high degree of accuracy, and in a fraction of the time which would be taken by older methods.

Finally, by way of rounding off this necessarily brief survey, it should be remarked that many of the devices used in industry are not, strictly speaking, measuring instruments, but should properly be described as comparators. That is, they do not measure the actual dimensions of an object, but simply compare it with a predetermined standard, in order to ascertain whether the dimensions fall within acceptable limits. Generally, these devices employ one or the other of the basic principles already described, but they are often designed with great ingenuity so that complicated measurements can be made to close limits by unskilled personnel, at high rates of output. In fact, appliances such as this are the very basis of modern mass-production methods; without them, the essential interchangeability of parts could never be guaranteed, no matter how complicated or efficient the production machines.

The measurement of time

We have already seen how the movement of the earth provided an absolute standard of time; its progress around the sun giving the duration of the year, and the time it takes to revolve upon its own axis, the day. We also saw how the early observatories were used to keep track of the one, and sundials the other. Until comparatively recently these methods, in refined forms, were still the most accurate means of time measurement, and the motion of the earth our absolute standard of time.

How Large is it?

In practice the sundial is rather inconvenient in northern countries, since the length of daylight in these latitudes varies with the season.* This, which is a matter of common experience, is due to the fact that the axis of the earth is inclined to the plane of its orbit around the sun. However, the principle of measuring the time of the earth's daily rotation remains valid, and it is only necessary to choose a reference point sufficiently far away. The modern astronomer makes this measurement by means of a 'transit telescope'. This is a telescope which is accurately aligned in the plane of the meridian, and equipped with cross-wires in the eyepiece. A suitable star is chosen as a reference point, and the time when it crosses the wires (the apparent motion of the star is due to the earth's rotation, of course) is noted. Twenty-four hours later, the earth's rotation will again bring the star to the cross-wires, and the interval of the two 'transits', being accurately recorded, can be used to check other time-measuring instruments.

It was most likely the relative sunlessness of northern climes which led to the invention of mechanical clocks. It can hardly have been the desire for greater accuracy, for the first ones were crude, and would have had to be corrected by the sundial at favourable opportunities. The early clocks, like the sundial, were only graduated in hours. There is an example of such a clock, dating back to 1386, in Salisbury cathedral, and this is believed to be the second oldest surviving mechanical clock—the oldest is in Paris. Another very old clock, still in working order, is preserved in the Science Museum, London. At one time this was thought to be even older than the Salisbury clock, but some doubt has been cast upon this. It would seem that the first clocks must have been made sometime in the thirteenth century. There had been other methods of time measurement, of which the water-clock, the hour-glass, and the graduated candle were probably the most used, but these are outside the main stream of development of time-measuring devices, and were rapidly supplanted once mechanical clocks became reliable, so they will not be treated here.

The primary requirement of a successful clock is some regularly recurring phenomenon which can be adapted to controlling the

* This would be equally true for southern latitudes, but it happens that the major land-masses, where civilization developed, are concentrated north of the equator.

motion of the clock. The early ones used a system of oscillating weights called a 'verge', but this was not very satisfactory. When, about the time of the Renaissance, the need for more accurate time measurement was felt, something better had to be found. The answer was the pendulum, the discovery of which is usually credited to Galileo, sometime around 1580. Legend has it that he was led to the discovery by watching a swinging lamp in the cathedral at Pisa. Timing the swings by means of his own pulse, he was able to show that they were isochronous, that is, the time taken for each swing is the same, irrespective of the length of swing. The time of swing is principally governed by the length of the pendulum; for an 'ideal' pendulum it can be calculated from the formula $T = 2\pi\sqrt{(l/g)}$, where T is the time, l the length and g is the acceleration due to gravity, the average value for which is $9 \cdot 80665 \mathrm{m.s^{-2}}$ (see also page 95). The exact formula for a real pendulum is rather more complicated, but the simple one gives results quite acceptable for many purposes.

Although Galileo realised that the pendulum could be applied to controlling a clock, he appears to have done nothing about it until almost the end of his life, when he drew up designs for a pendulum clock, which he instructed his son to build. However, the clock was never made, though the design was sound, and would have worked, and it was left to the Dutchman, Christian Huygens (1629–95), to develop the first practical pendulum clock. He also invented the balance spring, which made portable clocks and watches possible.

As exploration progressed, and long sea voyages became common, the need for a means of ascertaining longitude at sea became pressing. The most practical way of doing this is to compare the local time at noon, ascertained by observing the sun, with the actual time at some reference point on the earth's surface, which in practice is the meridian of Greenwich. This implies some sort of timekeeper which will show Greenwich time at all times and in all conditions, and the greatest error that can be tolerated for the purpose is 3 seconds per day. This problem was solved by John Harrison in 1762; his principal invention was the 'compensated balance' which cancels the adverse effects of varying temperature by making use of the differing expansions of two metals. From then until the early twentieth century, the story of timekeeping is largely one of refinements of mechanism. Unfortunately, however, the pendulum clock, and all other mechanical clocks, for that matter, suffer from certain disadvantages which are quite fundamental.

These disadvantages are of no account for everyday purposes; mechanical clocks do everything required of them by the man in the street, or even by the navigator, but for certain scientific purposes greater accuracy is desirable. To start with, the pendulum is not completely isochronous. Ways of overcoming this have been devised, but they necessarily add to the complication of the mechanism, and the same applies to the various methods used for correcting the inherent errors of the balance spring. Another difficulty is not so easy to deal with; a pendulum or balance can only be relied upon for perfect timekeeping if it is operating freely without any kind of outside interference. This is a condition which can only be secured for a few hours at most in practice, and would in any case be useless in clock-making. The function of the pendulum or balance is to control the rest of the clock mechanism, which actually indicates the time. In doing this it must of necessity interact with the rest of the mechanism, and the condition of no interference is at once violated. Furthermore, this interaction, though it is kept to the absolute minimum by ingenious design, necessarily robs the pendulum of some of its energy, so that, even if air resistance could be completely eliminated (by no means as simple as it sounds), it would eventually come to rest. Therefore a practical clock must include some means of giving the pendulum or balance a periodic impulse to keep it going, and this once more interferes with its free working.

Already by the middle of the nineteenth century attempts, more or less successful, had been made to provide the driving impulse electrically. But clocks built on this principle still need a switching device, and, moreover, they nearly all use the pendulum or balance to provide the driving power for the clock mechanism, rather than simply as a means of control, so they are not really any better as timekeepers than a well-designed spring- or weight-driven clock, though they are often more convenient. In passing, these true electric clocks should not be confused with the electric clock which is 'plugged in' to the domestic electricity supply. This is actually only an indicating device, not a timekeeper in its own right; the time is dictated to it by the frequency of the supply current.

Nevertheless, it was along the lines of electrical control that the next major advance was made. This was the invention, in 1922, of the *free pendulum* by W. H. Shortt. The Shortt free pendulum swings in a vacuum, under conditions of controlled temperature, and it does no work; the only outside interference with it is the giving of a small impulse every half-minute to keep it swinging. This impulse is applied at the most favourable part of the swing, so that its influence upon the timekeeping of the free pendulum is kept to a minimum. The timing of the impulses is looked after by a second, 'slave' pendulum, so called because the giving of the impulse is ingeniously utilised to keep the second pendulum in step with the free pendulum through a device called the 'hit and miss' synchroniser. Any impulses needed to operate external dials or timekeeping devices are taken from the 'slave' pendulum, and since the swings of this are constantly being corrected, the interference implied does not matter. The Shortt free pendulum can keep time to within about one second per year, and it rapidly became the standard timekeeper in observatories all over the world. It was also responsible, in Great Britain and else-where, for a much higher standard of public timekeeping, for free pendulums provided the control for the radio time signals which were (as they still are) broadcast at frequent intervals throughout the day, and they also made possible, indirectly, the exact standardisa-tion of the frequency of the public electricity supply, referred to above.

It was not long before even greater refinements became desirable, and now, the pendulum and the balance, which had served so long and so well, proved incapable of meeting the demands made of them. Some other isochronous process, with an inherently greater degree of accuracy, had to be found. Mention has already been made of the 'frequency' of an electric current. The currents flowing in the supply mains, and in a great many electrical circuits, as for instance, in a radio set, are not a steady, straightforward flow, of which a simple analogy would be the movement of water in a pipe, but they fluctuate, generally very rapidly, in some definite way, usually from zero to a maximum, then back to zero and again to a maximum, but in the other direction. Such a current, called an *alternating current*, flows, so to speak, backwards and forwards (it will be appreciated that these analogies are very crude; electrical phenomena should really be described in mathematical terms for the greatest precision).

Now, if an alternating current be applied to certain kinds of crystals, of which quartz is the most commonly used, the crystal will be set into vibration. Furthermore, the frequency of the vibrations is extremely constant, being determined by the internal structure of the crystal and the way it is cut. Also—and this is the important point— the frequency of the outgoing current will be the same as that of the natural frequency of the crystal—the current variations (technically termed *oscillations*) being, so to speak, 'pulled' into synchronism with the crystal vibrations.

An alternating current of this kind, vibrating at a very constant frequency, is extremely suitable for operating time indicators of a kind not greatly dissimilar to the familiar domestic mains clock. As a matter of fact, the natural frequency of the crystal is rather too high to enable it to be used directly in this way, and the output from the crystal is used to control a lower-frequency subsidiary current, which actually drives the clock, but the principle is unaffected. It would seem, then, that we have here the almost perfect timekeeper— small enough to be sealed in an evacuated container away from all external influences of temperature and atmospheric variations, with isochronous properties of a high order of accuracy, completely in-dependent of the amount of power applied or any other external influence, and providing an easily utilised output.

Unfortunately, this very perfection brought its own problems. It soon became clear that there was not very much point in correcting quartz crystal clocks by the earth's rotation, for they were accurate enough to time the movements of the heavenly bodies, which had for so long provided the ultimate standard of time, and showed it to be by no means uniform. The variations are by ordinary standards tiny, and for everyday purposes quite unimportant, but they exist, and are measurable. So the question arose, is there a standard time-keeper which can be used to check the performance of the quartz crystal itself?

The answer to this question is yes. For the ultimate standard of time, as of length, we have to go to the atom. In fact, the two are closely inter-related: this is implied by the statement on page 73 that a heated element gives out light of a very definite wavelength, which is only another way of saying, of a very definite frequency. This is a function of the very structure of the atom itself, and it provides our ultimate standard, but here the reader must be warned that any attempt at an analogy would be not only crude, but misleading. We are now dealing with something so far removed from our everyday world that any attempt to visualise it in any way of which our minds are capable is, practically by definition, incorrect.

The standard of timekeeping for scientific purposes is provided by the 'vibrations' (but remember the warning about analogies) of the atoms of the element caesium. A clock embodying this principle is accurate enough to tell us that the period of the earth's rotation (that is, the length of the day) may vary by as much as 0·025 second, the exact amount depending upon the season.

The measurement of power and energy

This is a comparatively recent development, and practically speaking, did not exist, because it was not needed, before the Industrial Revolution. But the extended use of machines, and more particularly of prime movers such as the steam engine, brought with it the need to measure and define such things as energy expended and work done.

Let us go back to first principles. For the present, we will use the English system of weights and measures. Suppose we have a weight, which we wish to lift through a definite distance. It is obvious that a certain amount of energy must be expended in doing this, and that this energy will be the same whether the job is done by human muscle power or by some machine such as a crane. It is equally obvious, and a matter of common experience, that more energy will be required to lift a larger weight, or the same weight through a larger distance.

So here is one way of measuring energy. We say that the energy expended in lifting a weight of one pound through one foot is one foot-pound. Equally, to lift a weight of one pound through two feet, or a weight of two pounds through one foot, is two foot-pounds, and so on. Going on a little, it should be fairly obvious that a weight of one pound *falling* through a distance of one foot also develops one foot-pound of energy, and perhaps a little less obviously, but reasonably enough when one thinks about it, a weight which has been lifted through a certain distance possesses an equivalent number of foot-pounds of *potential energy*, which will be converted to *kinetic energy* if the weight is allowed to fall.

Whenever energy is expended, work is done, and properly speaking, the foot-pound is a unit of work. Closely related to the ideas of work and energy is that of power, but to have meaning, this must possess a dimension of time. It does make a considerable difference, from a practical point of view, whether one digs the garden in one weekend or in six! The unit of power used almost universally for engineering purposes is the horsepower, which was introduced by James Watt in 1780 for the purpose of giving his customers some idea of what his steam engines (at the time a very new and strange invention) could do. Watt took as his unit the performance of a Greenock dray horse, and said, more or less arbitrarily it would seem, that this was equal to 33,000 foot-pounds per minute. As a matter of fact, only an extremely strong horse could keep up this level of performance for any length of time, so the performance of the engines would have been underestimated, probably no bad thing from the commercial point of view. But the original definition is still used.

Energy can exist in more than one form, and the various forms are interconvertible. One form of energy which is very important practically is heat. It has been shown by careful laboratory experiments that mechanical energy can be completely converted into heat energy. Unfortunately, the opposite is not true, and we cannot build a machine which will produce an amount of mechanical energy exactly equivalent to the heat energy put in. Partly this is due to the fact that no machine can be entirely efficient, but there are other, more complex reasons. To discuss these, however, would take us into the realm of thermodynamics, which is outside the field of this book. The point is, that it is important to be able to measure heat energy, and the unit in the English system is the British Thermal Unit (Btu). This is defined as the heat energy necessary to raise one pound of water through one degree Fahrenheit. Every user of gas has—or should have—some acquaintance with this unit, for gas, although measured by volume as a matter of convenience, is actually sold by its heat content, the unit for the purpose being the therm, which is 100,000Btu. From the definition given above, it is easy to calculate that a therm of gas should, in theory, boil about 65 gallons of water, starting at a temperature of 60° F—but remember that few domestic gas appliances are 100 per cent efficient! More will be said about the concepts of heat and temperature in Chapter 5; for the moment let us look at the units of work, power, and energy used for scientific purposes.

These units are based on the metric system, but the actual units differ according to whether the cgs or MKS system (see page 73) is in use. We will defer consideration of the unit of heat until Chapter 5. We start then, with the unit of *force*. In the English system, this is called the 'poundal', but it was not defined above, since it is not much used. The unit of force in the cgs system is the *dyne*, defined as that force which, if acting upon a mass of one gram, imparts to it an acceleration of one centimetre per second per second, mathematically, and much more conveniently, expressed as 1 g.cm.s^{-2} (acceleration being the rate of increase of velocity). In the MKS system, we substitute the appropriate units, giving the unit of force, the *newton*, as 1 kg.m.s^{-2}. Now, if the point of operation of a force of one dyne is allowed to move one centimetre in the direction of the force, work will be done, and this unit quantity of work or energy is called the *erg*. One erg is 1 g.cm^2.s^{-2}. The corresponding unit in the MKS system is the *joule*, which is 1 kg.m^2.s^{-2}. Only in the MKS system is there a named unit of power; it is the *watt*. Since power, as defined above, is the rate of doing work, it follows that 1 watt = 1 kg.m^2.s^{-3}.

None of these quantities can be measured directly. It is necessary to take measurements of all the individual quantities involved. Quantities like this, which are constant in any given case, but which may vary from case to case, are termed *parameters*. Thus, mass, length, and time are the parameters involved here, and given these, and the formulae set out above, the required quantities can be calculated. Let us take as a concrete example the measurement of the power developed by an engine or motor.

This is illustrated diagrammatically in Fig 11. Some form of brake is clamped to the flywheel or shaft, and attached to this is a lever, working between stops. A weight is adjusted along the lever until the latter is just held against the lower stop, meaning that the *torque*, or turning effort, of the engine is balanced by the weight. The power is thus acting upon the weight at the circumference of a circle of which the length of the lever is the radius. The relationship between the diameter of a circle and its circumference is known: it is represented by π (the Greek letter pi), and is 3·1416 to four places of decimals. We need now only count the number of revolutions which the engine

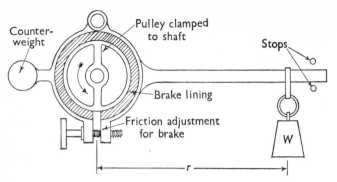

Fig 11   A 'Prony brake' for measuring the horsepower of an engine

is making in a given time (remember that power is the rate of doing work), and we have all the parameters needed to calculate the power output. Thus, if we wish to measure the power in terms of horsepower, application of the following formula will give us the answer:

$$\frac{Wr2\pi n}{33,000}$$

where W is the weight balanced, r is the length of the lever, and n is the number of revolutions per minute. The horsepower determined in this way is termed, for obvious reasons, *brake horsepower* (bhp). In Chapter 2 we saw how indicated horsepower, the power developed in the cylinder, is calculated. The ratio of one to the other gives the *mechanical efficiency* of the engine or motor concerned, a very important figure to the engineer.

Electrical units

It would be quite possible to fill an entire book with a description of the various measuring units and methods—in fact, it has been done—but we will bring this chapter to a close by considering the units used in electrical science and engineering, and sketching the methods of measurement. This is not the place to go into the question of what electricity is—a question to which in any case no simple or final answer is possible—but that it exists and can be used is a matter of everyday experience to everyone in a civilised country. To this extent, we are better prepared mentally than our ancestors, many of whom found great difficulty in grasping the idea of something which could not be seen, sensed (except in terms of an electric shock), weighed, or measured in any direct way. We, who are surprised when a lamp fails to light, or a motor to start when we turn the switch, have become used to the concept that things may exist in nature which can only be known to us by their effect on our instruments.

It is quite in order to talk of a quantity of electricity, and there is, in fact, a natural unit of this quantity—the smallest possible 'particle' of electricity which can exist—which is called an *electron*. Electrons are, incidentally, one type of the basic 'building blocks' of matter, and for this reason it is sometimes convenient, though far from accurate, to speak of them as particles. However, an electron is so unimaginably tiny that larger units are needed for practical purposes, and moreover, we are usually interested in electric currents, which may be thought of as the flow of numerous electrons.

We therefore start by defining a unit of electric current, and this, called the *ampere* (colloquially, 'amp'; abbreviation A), is defined in terms of one of the physical effects of an electric current. If such a current is passed through solutions of certain substances, it will break them up into their constituent elements (see Chapter 4 for a definition of the term 'element') and one ampere is that current which, when passed through a solution of silver nitrate in water, will deposit silver at the rate of 0·001118 gram per second. This is actually equivalent to a flow of $6·28 \times 10^{18}$ electrons per second, and this number of electrons, that is, the number passing a given point in one second when a current of one ampere is flowing, is the unit of electric quantity, or charge, called the *coulomb*.

Now there are several electrical quantities which can be measured, and they are all needed for some purposes, but we shall here deal with only the three most important ones, which are current, as noted above, and which is given the mathematical symbol I; resistance, which is the opposition to the flow of current, symbol R, and voltage, or more correctly, electromotive force (emf), which is the electrical 'pressure' driving the current, and the symbol for which is E. These quantities are related by the following rule, called Ohm's Law, after its discoverer, G. S. Ohm (1787–1854), and as it is basic to any understanding of electricity, it is worth while to commit it to memory:

$$R = \frac{E}{I}$$

which can also be expressed, according to the normal rules of algebra, as

$$I = \frac{E}{R} \quad \text{or} \quad E = IR$$

The unit of resistance is called the *ohm*, and has the symbol Ω (a Greek capital omega—the abbreviation O would be confusing when many figures might be involved). It is defined as the resistance, at $0°$ C, of a column of mercury 14·4521 grams in mass, of unvarying cross-section, and 106·3cm long, though this is nowadays only of academic interest.

Now that we have units for current and resistance, we can define a unit of emf by using Ohm's Law. This unit, called the *volt*, is the emf required to drive a current of 1A through a resistance of 1Ω.

Finally, multiplying voltage and amperage together, gives us a unit of electrical power, $IE = W$. This is the watt, and it is the same as the unit of power already defined in the MKS system, so we see that the electrical units can be linked to units of power and energy, and thus to the other basic units.*

The electrical properties mentioned above in the definitions of units would be somewhat inconvenient in practice, and it is necessary to make use of other phenomena in building practical measuring instruments. One possibility is to use the heating effect of an electric current. If a current is passed through a wire of high resistance, the wire will get hot. This is the principle of electric heaters, cookers, and other similar domestic appliances. Now if a substance is heated, it will expand, and the expansion will depend on the degree of heating. It can be shown that the degree of heating in the wire is directly proportional to the current flowing, so if we arrange to magnify the expansion of the wire (which is very small) by a suitable lever system, we can graduate a dial to read off the current directly. This instrument is called a hot-wire ammeter. It has some advantages, and finds

* It is therefore more logical to derive the electrical units from the basic units directly, and this is in fact done in practice, but the presentation adopted here makes for a simpler exposition.

a certain amount of use, particularly for measuring heavy currents, but for general purposes and especially for small currents, it is not very convenient.

Fortunately, there is another property of an electric current which can be used, and which is very well adapted to the purpose. Whenever an electric current flows in a wire, a magnetic *field* is set up around the wire, and if the wire is formed into a coil, this will behave as though it were a magnet, with its *poles* at either end of the coil. Most people are familiar with the properties of magnets; they attract iron and steel, and to a lesser extent cobalt and nickel; if a magnet be suspended so that it can swing freely, it will align itself roughly north and south; and the 'north-seeking' pole of one magnet will attract the opposite and repel the like pole of another magnet, and vice versa for the 'south-seeking' pole. If the reader is not acquainted with these facts, he may be interested to procure a couple of magnets ('bar' magnets for preference) and try them out for himself.

The strength of the magnetic field around a wire is exactly proportional to the current flowing in it, so we can make a current-measuring instrument (an *ammeter*) either by arranging for the current to flow through a fixed coil, and attract a piece of iron connected to a pointer—this is a *moving-iron* ammeter, a robust instrument suitable for measuring large currents—or by having it flow through a small coil of fine wire suspended freely between the poles of a powerful permanent magnet, and connected again to a pointer—a *moving-coil* ammeter, used where greater sensitivity is desired. For the utmost sensitivity, the pointer is dispensed with, and a tiny mirror is attached to the coil. This is arranged to reflect a light beam on to a scale, giving the effect of a weightless pointer, which can be as long as we like, within reason.

As stated above, instruments built upon these principles measure current, but if it is desired to measure voltage or resistance, it is quite easy to do so by making use of Ohm's Law. Thus, if a suitable known resistance is included in the circuit, the current flowing, and indicated by the instrument, will be proportional to the voltage applied, and the dial can be graduated to read directly in volts. In the same way, by using a definite known voltage, supplied by a suitable battery, resistance can be measured directly. It should be mentioned that there are other, intrinsically more accurate ways of measuring resistance, but the principles of these would take us rather more into the realm of electrical theory than is desirable in this book. These methods are in any case more suited to the laboratory than to the workshop, and much the same may be said of the many other ways of taking electrical measurements which can, in principle, be used. Nearly all the electrical instruments which the average person is likely to use or encounter work upon the principles just described— with one exception to be mentioned shortly.

The use of instruments of the kind described is not confined to electrical technology. Any quantity connected with a phenomenon which can produce or influence an electric current, can be measured by using such an instrument, and this is often more convenient than trying to measure it directly. We shall meet several instruments of this kind later; meanwhile, a good example is the exposure meter used by photographers. This contains a device called a photo-electric cell, which produces a small current varying according to the intensity of the light falling upon it. This current is applied to a very sensitive ammeter, which is graduated to read directly either in arbitrary light values or in camera settings.

Multiples and sub-multiples of the electrical units are named according to the usual rules of the metric system, but only the 'thousand' ones are much used in practice. Thus, for small measurements, we use milliamps, millivolts, and milliwatts. (Microamps and microvolts are recognised, but are more likely to be found in a scientific than in a technical context.) For larger measurements there are kilovolts and kilowatts, but currents of the order of thousands of amps are rare, so there is generally no use for a kiloamp. The abbreviations are mA, mV, and mW for the small, and kV and kW for the large. The abbreviations MV and MW may occasionally be met with; these should not be confused with mV and mW, as they stand for megavolts and megawatts, one million times the unit, respectively—voltages and powers which are most unlikely to be encountered in everyday life.

The instrument mentioned above, which does not work on the ammeter principle, is the ordinary electricity supply meter, and since this is such a common object nowadays, a brief description of its working may be of interest. In essentials, it is a very carefully and accurately built electric motor. It is so constructed that its speed, that is to say the number of revolutions which it makes in a given time, is proportional to the current passing. The supply voltage is constant, so this current is a measure of the power in watts ($IE = W$, page 98). This may vary from moment to moment, so, to make a workable unit of sale, the factor of time has to be introduced, and the unit by which electricity is sold is the kilowatt-hour. By means of suitable gearing, the revolutions of the motor are recorded on the dials of the meter, which are calibrated directly in kilowatt-hours.

4

What is it Made of?

As soon as Man had discovered the use of fire (and that is perhaps the oldest of all his discoveries) he must also have discovered that one substance could be changed into another, and it would have been a very short step from this to the realisation that all (or at least most) substances were made up of other, simpler substances. The study of these matters—what things are made of, and the changes which they undergo when mixed, heated, or otherwise treated, is the concern of the science of chemistry. Chemistry as a science is no more than about three hundred years old, but chemical processes have been used since before the dawn of history, and certainly long before anyone understood how they worked.

The first practical use made of chemical change was cooking. Brewing, tanning, dyeing, and the extraction of metals from their ores are other examples of chemical processes which were in use before the first civilisations came into existence. These are, incidentally, all quite complicated processes in the chemical sense.

The ancient Egyptians seem to have possessed a considerable body of practical chemical knowledge; thus the Ebers Medical Papyrus, which can be dated to the sixteenth century BC, shows that they knew how to prepare and purify a very wide range of drugs, while such things as glassware, dyed cloth, and cosmetics which have been recovered from the tombs prove that they must have had a considerable mastery of what we should now describe as chemical techniques. The Egyptians were, however, an empirical, pragmatic people, and appear to have been content to use their knowledge without speculating very much about the principles behind it. Not until the time of the Greeks do we find much in the way of chemical theory.

What is it Made of?

This is in many ways most unfortunate, for the Greek philosophers were by temperament and education quite unfitted for the study of chemistry. They held that all worth-while knowledge was discoverable by mental effort, and that practical work, which included experiment, was fit only for slaves. They developed geometry, which can be successfully studied along these lines, to a very advanced level, but chemistry is the experimental science *par excellence*, and it is therefore not surprising that most of the Greek thinking about it is pure speculation, and nonsensical speculation at that.

This might not have mattered much were it not that Greek thought and writing have influenced the western world ever since, and in particular their incorrect notions formed the basis of the operations of the first real experimental chemists, the alchemists. Alchemy has a very long history, reaching from late Greek times almost to the present day; indeed it is said that there are still people engaged upon the ancient quest. When the nature of that quest is explained, the reader may be tempted to the conclusion that the alchemists were foolish men who wasted their lives in the pursuit of a will-o'-the-wisp, so a careful explanation, with some qualifications, is called for.

The purpose of the alchemists was to find one or all of three things: the Philosophers' Stone, which would transmute the base metals into gold; the Elixir of Life, which would ensure to its possessor eternal youth, and the Alkahest, the Universal Solvent. We will take the last one first, since it gives the clearest insight into what was really going on. The great problem of the Universal Solvent would have been to keep it if it were ever found; in fact this is by definition impossible. We cannot really believe that the alchemists were so stupid as to have overlooked this snag, and we are forced to the conclusion that the Alkahest was a symbol for something else. The same applies to the Elixir of Life, and very largely, though not entirely, to the Philosophers' Stone. This explains the obscurity of much alchemical writing: it was not dealing with practical experimental chemistry at all. The meaning of the symbolism is something well outside the scope of this book; suffice it to say that many, perhaps most, of the alchemists during the Christian era were adherents of a mystical system which would most certainly have been regarded as heretical by the church.

There remains a body of men who were definitely pursuing an

103

experimental science, but it is not always easy to decide just what they were doing, because they also recorded their work in an elaborate symbolism, which cannot always be distinguished with certainty from that of the mystics. The reasons for this were twofold: first, just as the church frowned upon any attempt at free thought in matters of religion, it equally set its face against any attempts to find out the truth about the universe by the methods of experimental science, holding that all possible knowledge was contained in the scriptures. Under these circumstances it is not surprising that those who did try to find things out for themselves, and who most probably got some results very much at variance with accepted beliefs, found it politic to cloak their operations in secrecy. Secondly, anyone who really had discovered the secret of transmutation would have been in the greatest danger had the fact become known. The avaricious princes and nobles of the Middle Ages had ample means of compulsion at their disposal. Again, an involved symbolism, which would appear as nonsense to the uninitiated, was the best protection. It would appear that some of the alchemists managed to turn the greed and credulity of the nobles to their own advantage; many of them persuaded rich patrons to provide them with a comfortable living and a laboratory, in the hope that they would find the secret which would make their masters even richer.

Many of these latter alchemists were almost certainly real experimental chemists in the modern sense, pursuing knowledge for its own sake, using the golden bait as a way of getting the means to do so. This is amply proved by some of their discoveries, which could not conceivably have been made in the course of a genuine or simulated quest for the Stone. Since, however, they managed to keep their patrons reasonably happy, one suspects that a considerable knowledge of conjuring was not the least of their accomplishments.

Some of the alchemists stumbled upon useful information while genuinely seeking to transmute metals, and if we look at a case of this kind, it will serve to show how their minds worked, and at the same time will provide an introduction to the important concept of chemical change. It was widely held that the essential thing about a substance was its 'qualities', in other words, if one made something which had the outward attributes of gold, one would have made gold. Furthermore, it was considered that this result could be brought about by uniting substances with the desired attributes. Thus, gold is metallic, lustrous (it does not tarnish in air), heavy, and yellow. Mercury is metallic, lustrous, and heavy. Sulphur is yellow. Could we but combine them, so ran the argument, we should have gold.

More than one alchemist tried the experiment of grinding sulphur and mercury together. When this is done the result is not gold, nor is it anything like either sulphur or mercury. It is in fact a black powder, which is nowadays known as mercuric sulphide. This name, like all modern chemical names, has a definite meaning, and it tells us in fact what has happened in the experiment. The two substances have chemically *combined*, to form a new substance. It is important to notice that this chemical change, besides producing something quite different from the starting materials, is irreversible.

If we shake a mixture of oil and water violently, it will form a milky emulsion, which will separate into the original oil and water if it is allowed to stand. Similarly, water when heated turns into steam, which turns back into water when it is cooled down again. These changes, which can be reversed by altering the conditions of the experiment, are examples of physical changes. But no matter how long we let the mercuric sulphide stand, it will not change back into mercury and sulphur, nor will heating or cooling it make any difference. As a matter of fact, if it is very strongly heated, the mercury can be recovered in the metallic form, but this involves yet another chemical change, the other product of the experiment being a pungent-smelling gas. This is sulphur dioxide, formed when the sulphur combines with the oxygen of the air, leaving the mercury behind. A few chemical changes can be reversed by suitably altering the conditions, but the vast majority cannot.

Towards the end of the Middle Ages, the claims of the alchemists began to be discredited, though it was recognised that they had discovered a great many new and useful substances, and for the next couple of hundred years, this discovery of new substances, and in particular substances which would be useful in medicine, was held to be the true function of chemistry. This is known as the period of *iatrochemistry*.

During the sixteenth century, however, a new spirit had been sweeping across Europe; the spirit of the Renaissance and the Reformation. The mental tyranny so long exercised by the church was at last broken, and men began to feel free to think for themselves once more. During this period men like Galileo laid the foundations of modern science, firmly based upon experiment. Towards the end of the century, this spirit found its way into chemistry, and with the publication, in 1661, of Robert Boyle's famous book, *The Sceptical Chymist*, we can speak of the birth of chemistry as a science. This book defined for the first time the modern conception of an element. We shall have more to say about this later. About eighty years after this, Joseph Black introduced quantitative methods, which are fundamental to our modern conception of science, into chemistry, and these methods were soon brilliantly exploited by Antoine Lavoisier (1743-94) in his demonstration of the true nature of air and water and the phenomenon of combustion.

Lavoisier, at the height of his powers, was guillotined by the revolutionary innovators of the 'Age of Reason' for the most unreasonable crime of having been born an aristocrat, but he had pointed the way to one of the most important principles of chemistry, the Law of Conservation of Matter. This, in the form later codified and rigorously proved by experiment, states that in chemical processes matter can neither be created nor destroyed. In many ways, this principle is the key to all later development, which from this time on was continuous, and often spectacular. We shall now examine in rather more detail some of the things which were discovered by chemists in the course of this development.

MODERN CHEMISTRY

Today the body of chemical knowledge is so vast that it is well beyond the grasp of any single individual, and it is necessary to divide it up if any worth-while study, leading to new knowledge, is to be undertaken. There are several possible ways of doing this, but the commonest, and in many ways the most fundamental, division is into *organic* and *inorganic* chemistry. Organic chemistry is the study of the compounds of the element carbon; inorganic chemistry comprises all the rest of chemistry. The chemistry of carbon is the basis of life, and originally, organic chemistry meant the chemistry of living things, and the substances produced by them, but it has long been known that the processes of life are not different in kind from other chemical processes, as was once believed. However, the division is a convenient one, and it is retained with the definition given above.

It will now be necessary to explain two terms introduced in the preceding paragraph: *elements* and *compounds*. Once it has been realised that substances may be made up of other substances, it is at least a reasonable assumption that there must be certain fundamental substances, which can be put together to make up all the others. The Greek philosophers had held, following Aristotle, who lived from 384 to 322 BC, that all matter was made up of four ultimate substances: earth, air, fire, and water, which were called 'elements'. This theory was accepted for hundreds of years, and traces of it persist in our everyday language, when we refer to natural forces as 'the Elements'. It is not to be supposed that Aristotle was so naïve as to believe that all matter was composed of actual earth, air, fire, and water. We are rather to understand by these expressions the 'qualities' of dryness, fluidity, heat, and cold, in various proportions. Every substance was supposed to be composed of all the elements in various proportions, and to change one substance into another it was merely necessary to alter the proportion of the elements. This erroneous belief was the foundation of alchemy, and the latter did not really receive its death blow until Robert Boyle proposed the definition of an element which is accepted today.

According to this, an element is simply a substance which cannot be resolved into simpler substances by chemical means. The elements are thus, as it were, the fundamental building blocks, from which all else is constructed. This definition was an enormous step forward, and for chemical purposes it did not matter that many substances then classified as elements, because they could not at the time be broken up into simpler substances, were later resolved by more advanced techniques. Nowadays we have other sources of information to draw upon, and we can say with certainty whether a substance is an element or not. We thus know that there are at least one hundred and four of these natural building blocks, of which ninety are known to exist in nature. The others have been artificially created by atomic physicists, and do not exist naturally on the earth, though they may do so elsewhere in the universe.

All other substances are made up of these elements joined together (chemically combined) in various proportions, and are referred to as compounds. Some of the substances which we encounter in everyday life, including most of the common metals, are elements: examples are aluminium, copper, silver, gold, iron, and sulphur. Among other familiar substances salt, water, and sulphuric acid (the liquid in a car battery) are simple compounds, while sugar, cellulose (the basis of wood and many fabrics), and most dyestuffs are comparatively complex ones.

Various other substances are encountered which, though they consist of simpler substances, are not compounds, because the simpler constituents (some or all of which *may* be compounds) are not chemically combined; it would be possible, at least in principle, to separate them by purely physical (that is, non-chemical) means. These are termed mixtures: a mixture of salt and sand provides a good example; they can be separated by shaking with water, which is then carefully poured off and evaporated to recover the salt. Milk, wine, and air are everyday examples. It is sometimes extremely difficult to separate mixtures by non-chemical means, and chemical methods are often used in practice, but it can be done, and in a later chapter we shall see how the components of air are separated by purely physical means.

Elements do not combine to form compounds haphazardly, but only in definite proportions and according to definite rules. This observation led to the belief that there must be a certain minimum quantity of any element which could take part in a chemical *reaction*. This quantity is obviously very tiny; it is called an *atom*. As a matter of fact, we now know that individual atoms seldom take part in chemical reactions, and the smallest particle with which we normally have to deal in chemistry is called a *molecule*. This may consist of two or more atoms of an element linked together, or a number of atoms of different elements similarly linked. A molecule is thus the smallest particle of a compound which can exist. Even molecules are very tiny: the largest ones known, those of the complicated organic substances called proteins, are invisible in the most powerful of ordinary microscopes, and the simpler ones are unimaginably small. If a drop of water were magnified to the size of the earth, the individual water molecules would be about the size of tennis balls: a drop of water contains about 1.4×10^{21} molecules, and the diameter of a water molecule is approximately 2 Ångstrom units (page 74).

Chemists have a kind of shorthand for writing down the processes involved in chemical reactions, and though space in the present book is too limited to allow of its being described here, it is worth the while of anyone interested in the subject to master it. It can be studied in any textbook of elementary chemistry. Here it must be enough to say that every element has been assigned a symbol consisting of one or two letters, and these symbols can be combined to show the constitution of molecules of compounds, and they can also be used, according to certain definite rules, to illustrate the changes involved in a chemical reaction.

THE PERIODIC TABLE

We can, for the present, regard the atoms of any element as being identical, but different from those of other elements. It is a matter of common experience that a one-inch cube of lead weighs more than a similar cube of say, aluminium, and although we may not know the number of atoms in each, it is at least a reasonable guess that the atoms of lead weigh more than those of aluminium. In fact this is so, and their relative weights are one of the most important differences between the atoms of different elements. The technique for determining these *atomic weights*, as they are called, is a complicated one, and will not be described here. The point to bear in mind is that the atomic weight of any element is expressed as its ratio to that of a standard reference element. Originally this was hydrogen, the lightest of all the elements, but it is now found more convenient to take oxygen as the standard for chemical purposes, calling its weight 16. In physics, however, the atomic weights are related to that of the most abundant form of carbon. This double scale is in many ways a nuisance, but it need not detain us here.

Determining atomic weights was a long and laborious process, but by the middle of the nineteenth century the atomic weights of most of the 63 then-known elements had been pretty accurately established. It had also been realised that certain groups of elements exhibit striking similarities, and very often a sort of gradation of properties, which seem somehow to be connected with the atomic weight. Copper, silver, and gold are an example of one such group: their 'family resemblance' is well known to metal workers. The gradation of properties is not perhaps so obvious, but it does exist. Thus, copper dissolves fairly readily in strong acids, silver much less readily, and gold not at all. Attempts were made to explain these facts, but without much success until 1869, when the Russian chemist Dmitri Mendeleyef published his periodic table.

It is noticeable how, time and again, the same discovery has been made more or less simultaneously by different people working independently. Once the facts are known, it would seem, only the illuminating flash of genius is needed to show a new way of looking at them, and the discovery stands out as though by magic. This was the case with the periodic table, one of the most important discoveries in the whole of science. It was discovered by Mendeleyef, who worked from the facts of chemistry, and almost at the same time by the German, Lothar Meyer, who arrived at the same conclusions by studying the physical properties of the elements. Mendeleyef, however, published his results first, and his name is usually associated with the periodic table.

What Mendeleyef did, in essence, was to write out the list of the known elements in such a way that those with similar properties fell beneath one another, and then number them in order of their atomic weights, starting with hydrogen as 1. The elements then fell into fairly definite 'periods', but in order to preserve this arrangement it was necessary to leave gaps in the table. This might have worried a more pedestrian mind, but here we have an example of the inspiration of genius. Mendeleyef boldly asserted that the gaps were due to elements which had not yet been discovered, and he even went so far as to predict the properties of some of them. Thus he predicted the existence of an element in the same group as aluminium, which he called *eka*-aluminium (*eka* is Sanskrit for 'another'). Similarly he described an *eka*-boron and an *eka*-silicon. These apparently venturesome predictions were vindicated within a very short time. In 1875 a new element was discovered and found to possess the properties predicted for *eka*-aluminium. It was called gallium. *Eka*-boron and *eka*-silicon were discovered in 1879 and 1886 and named scandium and germanium respectively. In every case the agreement between the properties of the new elements and Mendeleyef's predictions was astonishingly close.

We can give just one more example of Mendeleyef's insight. There is a group of elements which was then called the 'rare earth' elements, and which are nowadays more usually referred to as the lanthanides, and all of which have very similar chemical properties. These did not fit at all well into the table at first, and Mendeleyef took what might seem to be the rather arbitrary decision to place them all in the same position. It was many years later, well into the present century, in fact, that the reason for the similarity of the lanthanides was finally explained, but here again, we now know that Mendeleyef was correct in his placing.

Gradually the gaps in the table were filled up, and its arrangement was refined. Today there are several ways of setting it out, each of which has its advantages and disadvantages. Fig 12 shows a somewhat simplified form, complete at the time of writing. The elements after actinium form a second 'rare earth' series, called the actinides. Most of them do not exist in nature, but are the products of nuclear physics. If there are any more elements, they must come after No 103, lawrencium*; we can confidently assert that all the lower places are filled. The chemical symbol for the element is given in each case, as well as its order in the list of elements, called the *atomic number*. This number is not just a convenient cataloguing device; it has a profound physical significance, into which we need not enter at present, however. The Groups, reading down the table, are elements with similar characteristics; in each case they show a distinct gradation of properties with increasing atomic weight. Group O, at the end in this table, is the 'rare gases' or 'noble gases'; they are extremely inert chemically and until recently were believed not to take part in chemical reactions at all. Just to show that chemistry still has surprises, however, compounds of krypton and xenon have been prepared within the last few years, and chemical theory has had to be revised accordingly.

---

* While this book was in preparation, the discovery of element No 104 was announced in Russia, and it was named kurchatovium. Little is known about it so far: it should in theory belong to Group IVA, and possess properties similar to those of hafnium and zirconium.

Periodic table — Groups

Period	I A	I B	II A	II B	III A	III B	IV A	IV B	V A	V B	VI A	VI B	VII A	VII B	VIII	O
1	1 Hydrogen															2 He Helium
2	3 Li Lithium		4 Be Beryllium		5 Boron		6 Carbon		7 Nitrogen		8 Oxygen		9 Fluorine			10 Ne Neon
3	11 Na Sodium		12 Mg Magnesium		13 Al Aluminium		14 Silicon		15 Phosphorus		16 Sulphur		17 Chlorine			18 Ar Argon
4	19 K Potassium	29 Cu Copper	20 Ca Calcium	30 Zn Zinc	21 Sc Scandium	31 Ga Gallium	22 Ti Titanium	32 Ge Germanium	23 Va Vanadium	33 As Arsenic	24 Cr Chromium	34 Se Selenium	25 Mn Manganese	35 Br Bromine	26 Fe Iron; 27 Co Cobalt; 28 Ni Nickel	36 Kr Krypton
5	37 Rb Rubidium	47 Ag Silver	38 Sr Strontium	48 Cd Cadmium	39 Y Yttrium	49 In Indium	40 Zr Zirconium	50 Sn Tin	41 Nb Niobium	51 Sb Antimony	42 Mo Molybdenum	52 Te Tellurium	43 Tc Technetium	53 I Iodine	44 Ru Ruthenium; 45 Rh Rhodium; 46 Pd Palladium	54 Xe Xenon
6	55 Cs Caesium	79 Au Gold	56 Ba Barium	80 Hg Mercury	57–71 *La–Lu	81 Tl Thallium	72 Hf Hafnium	82 Pb Lead	73 Ta Tantalum	83 Bi Bismuth	74 W Tungsten	84 Po Polonium	75 Re Rhenium	85 At Astatine	76 Os Osmium; 77 Ir Iridium; 78 Pt Platinum	86 Rn Radon
7	87 Fr Francium	111	88 Ra Radium	112	89–103 †Ac–Lw	113	104 Kurchatovium	114	105	115	106	116	107	117	108; 109; 110	118

* 57 La Lanthanum | 58 Ce Cerium | 59 Pr Praseodymium | 60 Nd Neodymium | 61 Pm Promethium | 62 Sm Samarium | 63 Eu Europium | 64 Gd Gadolinium | 65 Tb Terbium | 66 Dy Dysprosium | 67 Ho Holmium | 68 Er Erbium | 69 Tm Thulium | 70 Yb Ytterbium | 71 Lu Lutetium

† 89 Ac Actium | 90 Th Thorium | 91 Pa Protactinium | 92 U Uranium | 93 Np Neptunium | 94 Pu Plutonium | 95 Am Americium | 96 Cm Curium | 97 Bk Berkelium | 98 Cf Californium | 99 Es Einsteinium | 100 Fm Fermium | 101 Md Mendelevium | 102 No Nobelium | 103 Lw Lawrencium

Fig 12 The periodic table of the elements

What is it Made of?

CHEMICAL ANALYSIS

Broadly speaking, chemical operations are of two kinds, *analysis* and *synthesis*. Both words come from Greek: analysis means 'breaking down' and synthesis 'building up', and these terms explain the operations fairly well. Most modern chemical research is concerned with synthesis, but here, since we are concerned to find out what things are made of, we shall concentrate on analysis.

If we want to know what something is made of, we must break it up into its elements, and to make a complete analysis we should have to do this quite literally. For practical purposes, however, this is not always necessary: it is often sufficient for the purpose in hand if we can establish whether or not some element or group of elements is present. Possibly it may not even be necessary to know how much is present. For this purpose analytical *tests* have been devised. Certain reactions have been proved to be quite specific to various elements or groups of elements, and to many compounds also. We can thus rapidly 'test for' a given substance, or, by applying the tests in a systematic way, find out what an unknown substance is made of.

Chemistry is an experimental science: it is to be studied in the laboratory, not the library. The best way to get to know what chemical testing means, is to try it for oneself, and the reader is now invited to do this. Fortunately, for our modest purposes, a laboratory is not needed; any well-equipped kitchen will provide the facilities and most of the materials.

First of all, then, make a solution of starch by stirring a small quantity of the powder or lumps in hot water, and allowing it to cool. *Do not* use 'instant starch' for this experiment; it contains other things beside starch, which would invalidate your results by introducing unknown factors. This might not matter much for the present purpose, but it is as well to cultivate the habit of thinking scientifically right from the outset. Next we need a solution of iodine. The 'tincture of iodine' which is found in most first-aid kits is a solution of iodine in alcohol, and will do nicely, but if you have to buy something specially for the purpose, ask a pharmacist to make up a solution of iodine in potassium iodide. Now, to a small quantity of your starch solution, add a few drops of the iodine. The mixture will instantly turn an intense blue colour. This test is specific for starch, but you ought not to take my word for it. Two common substances closely related to starch are sugar and cellulose: try the same experiment with these, and note that there is no colouration. Cotton is a very pure form of cellulose, and can be used for this experiment, but make sure it is well washed first to remove any starch which may be present.

Having assured yourself that the test is indeed specific for starch, you can use it to find out whether various items contain starch. Try it first of all on a cut slice of potato. The blue colour here is quite strong, indicating that a potato contains a great deal of starch (that is why it is avoided by people who are trying to reduce their weight). If you have a microscope, even a very small and simple one, you can see that the starch is actually present in the form of tiny grains—these are stained blue, while the rest of the tissue is unaffected.

An experienced chemist learns to recognise many substances by their properties without necessarily having to apply rigorous tests, and this ability is very useful when conducting tests for more complicated substances. For your next experiment, take a wide-mouthed glass jar—a jam jar or pickle jar does well—and place in it a layer of common washing soda about half an inch or so deep. Then add sufficient vinegar to well cover the soda. A violent effervescence begins at once, indicating that a gas is being given off. Wait for a short time for this gas to push the air out of the jar, then investigate its properties. Lower a lighted match into the jar. It goes out at once —so the gas is not air. It has no taste or smell, but you will not be able to verify this because the smell of the vinegar would mask other odours. As a matter of fact, the vinegar is a convenient source of acid—acetic acid in this case—but almost any other acid would do, and if you have one to hand, which has no smell, you could use that instead. Dilute sulphuric acid (commonly known as 'accumulator' or 'battery' acid) would do well, but it is not a good idea to bring this into a kitchen, so if you use it work in an outhouse or shed, and take great care not to spill it, on yourself or anything else. Tartaric acid can also be used, and may be found in some kitchens, though it seems to have gone out of fashion these days. To put the matter beyond doubt, however, you should collect a sample of the gas uncontaminated by the vinegar or other acid, and this will anyway be needed for the next test.

To collect the gas, use a jar—or a bottle—with a well-fitting cork as the reaction vessel. A piece of tubing is needed to carry the gas away. Glass tubing is always used in the laboratory, but for our experiment it will be easier to use a piece of plastics tubing, about $\frac{1}{4}$in diameter and a foot or so long. This is easy to obtain, and does not require any skill in glassworking. Make a hole in the cork, of such a size that the tube can just be pushed tightly into it. A round file, pushed through the cork with a twisting motion, will do this job quite efficiently. Arrange the other end of the tube to dip under water contained in a bowl or basin. Now take a second jar—the kind in which cocktail olives and similar items are sold is ideal for the purpose—fill it completely with water, and keeping the mouth of it closed, either with the screw cap or with your hand, carefully invert it with its mouth under the water in the bowl. The mouth may now be uncovered, and the water will remain in the jar, held there by the pressure of the air.

Put the washing soda and vinegar or other acid in the reaction jar as instructed above, and then insert the cork, with its tube, firmly. **Warning.** Never close any vessel, in which gas is being evolved, with a solid cork. In the present case the pressure generated will certainly be enough to blow out the cork with great violence, and it might even shatter the jar—ordinary commercial glass jars sometimes have unsuspected weak spots. The cork with the tube is quite safe, however, and gas will pass down the tube and begin to bubble out under the surface of the water. Allow some time, as before, for the air to be displaced, and then bring the second, water-filled jar over the end of the tube. Gas will then rise into the jar, displacing the water, and when it is full (it will of course appear empty, since the gas is invisible) it can be closed and removed.

Try the test with the burning match again. The gas does not support combustion, and equally obviously, it does not itself burn. What else can we find out about it? It is heavier than air. One way of showing this is to let a jar of it stand with its mouth open for a short time. A lighted match plunged into it will still be extinguished, showing that the gas has not escaped, as it would do if it were lighter than air. There is a very elegant experiment for demonstrating this, which, though it requires some manipulative ability, is well worth trying. Fill a jar with gas, as already described, and use as large a jar as you conveniently can. Now get a short piece of candle, not more than an inch long. You will stand a better chance of success if you use one of the small candles used for Christmas decorations and similar purposes. Light this, and stand it on the bench or table. It is essential that the experiment be carried out in a room absolutely free of draughts. Now open the jar of gas, and, holding it about six inches or so above the candle, tilt it so as to pour its invisible contents on to the candle. If all goes well, the candle will be extinguished as though you had poured water on to it, thus proving that the gas is much heavier than air. The gas concerned is, in fact, carbon dioxide, and it is actually more than $1\frac{1}{2}$ times as heavy as air.

The properties described are alone enough to raise a reasonable presumption that any gas possessing them is carbon dioxide, but the matter can be settled by bubbling the gas through, or shaking it with, lime water. This is made by shaking some slaked lime (hydrated lime —the stuff one puts on the garden) with water and pouring off the clear fluid. In the presence of carbon dioxide this clear fluid goes milky, as you can easily demonstrate with the apparatus which you now have. It is also possible to show, given sufficient patience, that exhaled human breath contains carbon dioxide, by blowing bubbles through a suitable tube into a jar of lime water.

One important property of substances, which we often need to determine in chemistry, is their degree of acidity or alkalinity. Once again, a definition of the terms would take up too much space, and the reader must be referred to a suitable textbook, but the method of testing, in its simplest form, is easy to describe and to perform. It depends upon the fact that there exist a large number of substances, known as *indicators*, which show a change of colour in acid or alkaline conditions. Several of the substances commonly found in households have this property. Cochineal—used in cookery as a colouring agent—is one such, and various plant and fruit juices also give results. A solution made by boiling a few leaves of the common red cabbage in water does well. You can easily demonstrate the colour change by using 'white' vinegar as an acid and washing soda or ammonia solution as an alkali, and you can then go on to discover which common substances are acids, which alkalies, and which are *neutral*—that is, they have no effect on the indicator. It is also possible to show by this means how an acid is neutralised by an alkali, and vice versa.

These simple tests are not often used in laboratories nowadays. More precise determinations are usually required, and the exact degree of acidity and alkalinity must often be ascertained. At one time, this was determined by titration, which works roughly as follows. First of all, *standard solutions* of acids and alkalies are made up. The particular acids and alkalies used for the purpose must be carefully chosen; they must be extremely pure. A known weight of the material is then dissolved in a known volume of water, using special techniques to ensure that all the material is in fact present in the solution.

We will suppose that we have a standard acid solution, and that we wish to determine the strength of an alkaline solution. Using a pipette (descriptions of the measuring instruments were given in Chapter 3), we measure a definite quantity of the 'unknown' solution to be tested into a suitable vessel, usually a flask. A burette is filled with the standard acid solution, which is adjusted accurately to the zero mark. Now a few drops of a suitable indicator are added to the unknown solution—the exact one used will depend to some extent on the nature of the acid and alkali used; one substance very commonly used is methyl orange, which would make the liquid straw-coloured. The standard solution is then run into the unknown solution from the burette, a little at a time, the flask being shaken after each addition. At a certain point the indicator will change colour to orange-pink, showing that the alkaline solution has been exactly neutralised. In practice, a rough titration would be made first, to find the approximate 'end-point', and then a more accurate one would be made, adding the acid only one drop at a time as the end-point was approached. The amount of acid needed to neutralise the unknown solution can then be read off from the burette. Again, in a practical test, at least three titrations would be made, and the average result would be taken—this is to minimise the inevitable experimental error.

Now, we know that the substances taking part in a chemical reaction do so only according to definite rules and in definite proportions, which are expressed by small whole numbers. Moreover, any known reaction can be written down in shorthand form as a *chemical equation,* as mentioned on page 109. We therefore look up in a reference book the equation for the reaction between the acid and the alkali with which we have been working, and then, knowing the amount of acid used, it is a matter of simple arithmetic to calculate the amount of alkali present.

This technique is *volumetric analysis.* It is not confined to the determination of acids and alkalies, but can in principle be used to determine the quantity of any element or *radical* (a group of elements in combination, which acts rather like a single element) present in a given compound, provided that: (a) the substance concerned can be dissolved in some suitable solvent; (b) that there is another substance, which will react in solution with the first one, and which can be prepared in an accurate standard solution; (c) that it is possible to follow the progress of the reaction to a definite 'end-point', either by changes in the properties of the reaction mixture itself, or by means of some suitable indicator. Thus, the starch-iodine reaction can be quantitatively followed in this way, and many other volumetric tests are known, and are used in laboratories every day.

Titration is not much used nowadays as a measurement of acidity or alkalinity, however, since a much quicker and more accurate method is available. The degree of acidity can be expressed by the *pH scale.* The pH of a solution is a measure of the *hydrogen ion concentration.* The meaning of this, and the methods of measuring it, are the subjects of whole, and by no means simple, books, so space prevents us going deeply into the matter here. Suffice it to say that a pH of 7 represents a neutral solution: less than 7 is acid, and the smaller the number the stronger the acid; very strong acids have minus numbers. Conversely, a pH of more than 7 represents an alkaline solution, the higher the number the stronger the alkali.

All modern laboratories have some kind of pH meter, most of which depend upon electro-chemical principles. If two different, suitably chosen metals are immersed in an acid solution, and connected by a wire, an electric current will flow between them; this is the general principle of the electric cell. In a pH meter the solution to be tested is made to act as half of such a cell, the other half being a 'reference electrode' working under known conditions. Under these circumstances, matters can be arranged so that the current which flows when a test electrode is placed in the test solution is exactly proportional to the pH. This current is applied to an ammeter, as described in Chapter 3, but the instrument is graduated so that the pH can be read off directly on the scale, and no calculation is needed; the test can be completed in seconds.

Gravimetric analysis

The inherent accuracy of the weighing process was pointed out in Chapter 3. For the most accurate quantitative work, therefore, we turn to weighing, and this method is known as gravimetric analysis. As with all other analytical methods, we must find a reaction suited to the purpose in view, and we must be able to write, or to look up, an equation for the reaction. To illustrate the methods, let us suppose that we have been handed a mixed lot of metal scrap, which is known to contain some silver. We are required to find out how much. In this case, mainly for the sake of simplicity, it will be assumed that the other metals can be disregarded, also that neither lead nor mercury is present. In a practical situation this might well not be so, and the procedure might then be different, and would certainly be more complicated, but the principles would be much the same.

First of all, then, we dissolve a carefully weighed quantity of the mixed metals in nitric acid, which is a strong acid known to dissolve silver completely, forming a compound called silver nitrate. For the present purpose, it does not matter whether the other metals dissolve or not; if any are left after heating the solution and allowing long enough for the silver to dissolve completely (we naturally use plenty of acid to make sure of this), they can be filtered off. So now we have a solution of silver nitrate, possibly mixed with the nitrates of other metals. Our problem is to recover the silver from this in a form which can be weighed.

Note that in this case it is not necessary to isolate the pure metal, as we might sometimes have to do, and this simplifies matters considerably. We have to get the silver out of solution, and fortunately, it is a general principle in chemistry that if two substances in solution can react to form an insoluble substance they will do so. There are very few exceptions to this rule. We take a solution of sodium chloride, which is common salt, and which is known, from experiment, to react with silver nitrate, forming silver chloride and sodium nitrate, but not with any of the other metals which might be present.

Silver chloride is insoluble, and when the two liquids are mixed, it appears as a dense white *precipitate*. This has now to be separated from the remaining liquid, which is done by filtering. The whole of the reaction mixture is poured on to a special paper, not unlike blotting paper, which is formed into a cone shape and contained in a funnel. The liquid runs through the paper, which is porous, and the precipitate remains on the surface. The process is generally hastened by applying suction. The reaction vessel is carefully washed out, and the washings are added to the filter, to make sure that all the precipitate is recovered, and to ensure that it is not contaminated by soluble impurities, the precipitate is washed several times with water while it lies on the filter.

It is now necessary to weigh the silver chloride. This cannot be done directly by removing it from the filter, since there would be no hope of getting it all off without taking some of the paper with it; equally we cannot weigh it on the paper, since the weight of the latter may vary considerably due to contained moisture and other factors. What is done in fact is to dry the filter paper with the precipitate very thoroughly, and then burn it to ash in a crucible. The ash and the precipitate can be weighed, and if the weight of the ash from another, unused filter paper is subtracted, the result will be the weight of silver chloride. Filter papers are specially made so that the amount of ash which they leave on burning is very constant. Now, the composition of silver chloride is known: it contains 75·3 per cent silver and 24·7 per cent chlorine, by weight, and nothing else. It is thus a very simple matter to work out how much silver is present in a given weight of silver chloride. Knowing the weight of the sample we took, and assuming the composition of the bulk material to be the same, an ordinary sum in proportion will tell us the amount of silver in the sample given to us, which is what we wanted to know.

Organic analysis

A knowledge of organic chemistry is not necessary to the understanding of any of the other techniques described in this book, and the subject, which is a fascinating one, really requires a book to itself to do it justice. Fortunately, some good popular introductions are available, and are mentioned in the Reading List. We will, however, just look at one method of organic analysis, mainly because it illustrates so well the meaning of the word analysis.

Most organic substances have very complicated molecules; it is not that they necessarily contain a great many different elements, but a single molecule may contain a very large number of atoms of the same elements, the arrangement of which is of great importance. In an organic analysis the arrangement, as well as the proportions, of the constituent atoms, must be ascertained. Nowadays spectroscopy (see page 128) is a great aid in all the processes of organic analysis, but the older method will be described here.

The first thing which an organic chemist, faced with an unknown substance, must do, is to find out which elements it does in fact contain. This is a fairly simple matter; methods of qualitative analysis similar to those already described are used. Organic substances all contain carbon (this by definition—see page 106), and nearly all contain hydrogen. These elements can be detected by heating the substance with something which will provide oxygen; copper oxide is generally used. This oxygen combines with the carbon, giving carbon dioxide, which can be detected with the aid of lime-water (see page 116), and with the hydrogen, giving water. There is a specific test for water: if some copper sulphate crystals are strongly heated, they will turn white. Water will restore their blue colour, and it is the only liquid which will do so.

Various other substances are 'tested for' in the same way, and when the chemist is sure that he has accounted for all the constituents, he can determine the percentage of each which is present by the general techniques of gravimetric analysis already described. This, however, is only half the story! To take an actual example, a certain compound may be shown to contain 40 per cent carbon, 6·7 per cent hydrogen, and 53·3 per cent oxygen by weight. This corresponds to a proportion of atoms in the molecule of 1:2:1 carbon, hydrogen, and oxygen. If the actual number of atoms is the same as this proportion, the substance is formaldehyde, widely used as a disinfectant. But this proportion can equally well represent an actual composition of 2:4:2—in which case the substance might be acetic acid, the acid in vinegar—or any other numbers in the same ratio, such as 6:12:6, the composition of a number of sugars, one of which is glucose.

121

It is therefore necessary to find out the actual size of the molecule, that is, the *molecular weight* of the substance. To do this, we make use of the fact that a certain number of molecules, no matter of what kind, will lower the freezing point of water by a definite amount. So if we take a carefully weighed amount of the unknown substance, dissolve it in a known quantity of water, and accurately determine the freezing point of the solution, we can calculate the number of molecules in the sample, and hence the molecular weight. If in our example, the molecular weight shows the compound to have the simplest composition mentioned (which in chemical terms is CH_2O) then the matter is settled; it is formaldehyde, since there is only one possible arrangement of atoms corresponding to this *formula*. But there are three possible arrangements for the second one mentioned ($C_2H_4O_2$), and no less than twenty-two for the formula $C_6H_{12}O_6$. It is therefore necessary to determine the arrangement before the substance can be named with certainty.

This is a matter of literal analysis. The compound is actually split apart, or broken down, to give simpler constituents which are identified one at a time. Then the manner in which the constituents fit together must be deduced, and finally a structure is drawn up which satisfies all the evidence. This may be a very long and tedious process, extending even over years in the case of a complicated molecule. Finally, if the substance is a naturally occurring one, it is not generally considered that the suggested structure is proved until it has been synthesised—that is, built up from simpler substances by definitely known processes, and this may very well take as long as, or longer than, the original analysis.

MICROCHEMISTRY

Modern chemistry is, to an increasing extent, concerned with the handling, testing, and detection of very small amounts of substances. To some degree, this simply reflects the advancement of the science of chemistry itself; it was only natural that the earliest substances studied should have been those most abundant. Apart from this, however, it happens also that many interesting processes, particularly in biochemistry, the chemistry of living things, are concerned only with minute amounts of material. Furthermore, radiochemistry, that is, the chemistry of the radioactive elements, has also to deal with very small quantities, partly on account of the actual rarity of the substances concerned, and partly because they are extremely dangerous, and the less radioactive material in a laboratory at any one time, the better. The story has often been told of how the entire chemistry of the element plutonium, one of the 'man-made' elements, was worked out at the UK Atomic Energy Research Establishment at Harwell with a sample about the size of a pin's head.

In essentials, the methods used for analysing tiny quantities are simply scaled-down versions of those used for larger quantities. In the early days, in fact, miniature versions of the standard laboratory apparatus were used—and still are for some purposes, today, but it soon became obvious that some modifications to the technique were essential if microchemistry, as it came to be called, were to be fully developed. To begin with, small traces of impurities, unimportant on a larger scale, could completely invalidate the results of a micro-analysis. It is obvious that very pure reagents are essential, but some other sources of error are not so obvious. For instance, rubber tubing is widely used to couple up the glass tubing which forms an important part of most equipment. One of the pioneers of micro-chemistry found that small amounts of volatile substances, sufficient to upset his results, were coming from the rubber tubing, which for microchemical work has to be specially treated to eliminate this source of error.

With these techniques, successful analyses can be made with only about 20mg of sample. Success led to even greater refinements, and the methods used became more 'direct'. Thus, analyses are often made on a microscope stage, using only a drop of solution; there is no containing vessel in the ordinary sense. A whole range of tests and titrations can be made using tiny burettes made from extremely fine glass tubing; the general principles, however, remain the same as for larger-scale work. These techniques have become known as 'ultra-micro analysis'—this term, like 'microchemistry', is sometimes objected to by purists, but both have become established. Of course, the most important advance, which has virtually made these techniques possible, is the development of the extremely accurate balances mentioned in Chapter 3. This will be understood if it is mentioned that techniques are now being actively developed which can operate successfully with as little as 20–50 micrograms of sample (a microgram is 10^{-6} gram).

One technique of microchemistry which deserves special mention is the ring-oven method, notable as much for its ingenuity and simplicity as for its great usefulness. Using this method, amounts of metal, for example, of less than one microgram, can be separated. The apparatus for this method contrasts sharply with the elaboration and complexity of a good deal of chemical equipment, consisting only of a solid metal cylinder with a hole drilled axially through it. This is arranged so that it can be evenly heated, and a pipette is clamped over the centre of the hole. In use, a drop of the solution to be tested is placed in the centre of a filter paper, which is then placed over the cylinder, and washed with a suitable solvent delivered from the pipette. The solvent spreads outwards from the centre just as a drop of water placed on blotting paper does, and it carries the test material with it in solution. At the periphery of the central hole, when it comes into contact with the heated ring, the solvent evaporates, leaving the test solution concentrated into a ring. The filter paper is then cut up into segments, which are tested for various elements and radicals, the procedure being in general much the same as that already described for larger-scale work. Much use is made of tests which employ colour-changes, of course. In this way, a very large number of tests can be made on a very small sample, and the technique is especially useful for dealing with rare and valuable substances. It has been used, for example, to determine the composition of ancient coins, bronzes from the Egyptian tombs, and the pigments in old paintings. We shall meet these applications again in a later chapter.

CHROMATOGRAPHY

The process just described has some features in common with one of the most useful of all techniques for detecting small quantities of material—paper partition chromatography. Before we look at this, however, we will consider chromatography in general, and the principles on which it is based. It has already been mentioned, when discussing the difference between mixtures and compounds, that the separation of mixtures is often a very difficult matter. On the other hand, many naturally occurring materials, concerning which we would like to know more, are mixtures, and chromatography is one of the most valuable tools of modern research because it offers a ready means of rapidly separating mixtures into their constituents.

Biochemists have always been interested in the isolation and identification of minute quantities of material, so it is not surprising that much of the pioneer work in chromatography was done by these workers. The method was discovered in 1906 by Michael Tswett. The underlying principle is very simple. Suppose we have two immiscible substances, for example oil and water, contained in a suitable vessel. This set-up is called a *system*, and the constituent substances are termed *phases*. If now a mixture of two substances is introduced into this system, we shall find, generally speaking, that these substances distribute themselves in unequal concentrations between the two phases. This will be clear if we take a very extreme example, such as a mixture of wax and salt. In the oil/water system, this mixture would be separated completely, the salt dissolving in the water, and the wax in the oil.

Even when the difference is not as great as this, it is still possible to achieve complete separation by repeating the process many times, and chromatography is, in effect, a means for doing this continuously. This is done by holding one of the phases fixed in a *column*, which is usually a glass tube about 3cm in diameter, and passing the other phase through it. Tswett's method (still much used) was to have the column packed with a solid adsorbent, ie a substance which absorbs various substances on its surface, and a liquid solvent flowing through it. Activated alumina (a specially purified form of aluminium oxide) and benzine form a widely-used combination. Tswett used this method to separate the pigments of green leaves, and noticed that the various pigments appeared as differently-coloured bands on the column. He therefore called the method chromatography, a name which is still used today, even though colourless substances are now more often investigated.

It is fairly clear from the above description that the effect of the unequal distribution of the constituents of the mixture (there may be more than two, of course) between the phases is to cause them to be held in the column for varying lengths of time. Thus, given a suitable method of detection, the various constituents may be collected separately as they reach the bottom of the column. They may be

125

detected by colour, as in the original work, or by any other suitable chemical or physical characteristic. For example, some substances fluoresce, that is, they glow, in ultraviolet light. If the column be irradiated by an ultraviolet lamp, such a substance will be immediately revealed.

The next step forward was suggested by A. J. P. Martin and R. L. M. Synge, in 1941, and actually developed by Martin and James in 1952; it is thus a fairly recent technique. This is gas chromatography. Here, the liquid solvent is replaced by an inert *carrier gas*; the other phase may be solid, as before, or liquid, which is rather more usual. If it is a liquid, it is supported in the column by an inert material, crushed brick being quite a usual one. The column in this case is 2m long and 4mm internal diameter, formed into a U-shape for convenience. These dimensions are standardised. There is a wide choice of liquids for the liquid phase; a common one is liquid paraffin.

In use, gas from a cylinder is passed through a flow controller, and then through a sample-injection system, which is a device enabling the sample to be injected into the gas stream without interrupting the flow. The gas, now carrying the substances to be separated and tested, then passes to the column. The method is easiest to operate with fairly volatile substances, but materials of higher boiling points can be dealt with by heating the column; its temperature must in that case be maintained constant. At the far end of the column is the detector, which, as its name implies, detects the appearance of the various constituents in the gas stream leaving the column. There are several forms of detector; one of the most popular depends upon the fact that the thermal conductivity of the gas varies according to its constitution. This property is fairly easy to measure, and it can be shown directly on a dial or recorded by a pen on a continuously moving paper chart. In the latter case the pen-trace will show a series of peaks, corresponding to the times when the various constituents leave the column.

This is an important point. For a given set of conditions (gas, liquid, column size, temperature, rate of flow, etc) the *retention time* of a given substance on the column will be the same. We can thus identify the constituents of the original sample, either by passing a number of pure samples of known substances through the column, and noting their retention times, or, much more conveniently, by comparing the recorded traces with published data collected under the same standardised conditions. Furthermore, the *size*, as distinct from the position, of the peaks has been shown to correspond to the amount of substance present, so we have here a means of quantitative analysis as well. Properly carried out, this technique is very accurate, as well as being extremely sensitive, and it is in fact the method used for assessing the amount of alcohol in the blood of a motorist who has failed a 'breathalyser' test. This calls for an accuracy of something like $\pm$ 2 parts in 100,000, and the sample may consist of no more than a few drops of blood.

We can now turn, with better understanding, to the method of paper partition chromatography mentioned earlier. The underlying principle here is rather different, but still very simple to understand: it depends on the fact that different compounds are, in general, soluble to differing degrees in the same solvent. To take a homely example, you can dissolve a pound of sugar in a pint of water by heating it (often done in fruit-preserving for instance), but a pound of salt will not dissolve in this quantity of water, even if it is heated to boiling.

The apparatus for paper chromatography is also very simple: a sheet of filter paper and a vessel to contain the solvent. The unknown mixture to be tested must either be a liquid already, or a suitable solvent which will dissolve all its constituents must be found by experiment. This is not usually too difficult. A drop of the mixture is then 'spotted' on to the filter paper near one edge, together, perhaps, with samples of known compounds suspected to be present in the mixture. The paper is then stood on this edge, dipping into a shallow layer of a suitable solvent, which may itself be a mixture. The solvent is soaked up by the paper, and gradually passes to the opposite edge, passing over the sample spot as it does so. As the various constituents of the spot are dissolved, they are separated, so that each travels across the paper at a different rate. This process is called development; it is complete when the 'solvent front' has reached the opposite side of the paper. The paper is then dried, and sprayed with a reagent, or a mixture of reagents, chosen so that the various constituents of the sample will show up as coloured spots. The colours serve to identify the compounds; an additional check is

127

provided by the distances travelled, which are specific for each compound, and which can be checked by reference to the known compounds which were included. If the separation of the various constituents is not sufficient at the first attempt, a 'two-way' technique is adopted. The test solution is spotted on the paper near one corner. The *chromatogram* is then developed as before, when the constituent compounds will be spread out along one edge. The paper is then turned at right-angles and the process repeated. In this way, the various spots are spread out across the paper, and identification is now an easy matter.

As will be realised, this technique is very sensitive, and amounts of only a few micrograms of material can be detected. It has been widely used for identifying the composition of materials produced in tiny amounts in biochemical processes—for instance, in a living leaf exposed to sunlight for only a few seconds. In this way, the progress of the photosynthetic process, which is the basis of all life on earth, can be followed, and much of our present understanding of this process depends on paper chromatography.

SPECTROSCOPY

The oldest method for the detection of small quantities of matter, and still the most useful one, is spectroscopy. It has already been pointed out, on page 73, that a heated element gives out light of definite wavelengths. In practice, the wavelength of light is recorded by our eyes as colour. 'White light' is a mixture of wavelengths. This was proved by Isaac Newton in 1670, when he passed a beam of sunlight through a prism (a block of glass of triangular section) and noted that it was converted into a broad band of coloured light, which could be projected on to a screen. The light passing through the prism is refracted—the phenomenon of refraction is explained in Chapter 6—but the various wavelengths are refracted to different degrees, and so the white light is split up into its constituent colours. The coloured band so obtained is called the *spectrum*; the colours merge imperceptibly into one another, but six principal ones can be distinguished: red, orange, yellow, green, blue, and violet—the colours of the rainbow.

These facts were well known by 1852, when Robert Bunsen took up a professorship of chemistry in the University of Heidelberg, and there met the physicist G. R. Kirchhoff, so beginning one of the most fruitful partnerships in the history of science. This unlikely pair—they were quite different in habits, temperament, and even physical appearance—were firm friends even outside the laboratory, and worked together to such effect that by 1859 they had laid the foundations of the science of spectroscopy. Each man was a great discoverer and innovator in his own right: Bunsen's name, in particular, is, even to the most unscientific, virtually symbolic of science, because of the gas burner which he invented, and which is found in every laboratory in the world. Kirchhoff also did much independent work, particularly in the field of electricity.

Bunsen and Kirchhoff reasoned rather as follows: we know that many elements, when heated, give out light of a very definite colour, that is to say, of only one, or at most, only a few wavelengths. Thus, sodium or one of its compounds, if heated in a colourless gas flame, produces a striking yellow colour; potassium, a closely related element, has a violet flame, and so on. If we could separate these wavelengths, and record and measure them in some way, we should have ready means of identifying an element, even among a mixture of others.

The instrument which they devised for doing this is the spectroscope. Probably it was mainly the work of Kirchhoff, as the physicist of the partnership, though it is never possible to be sure how much individuals contribute in cases like this. It is on record that the first one was made from a prism, a cigar box, and a couple of superannuated telescopes! The modern form of the instrument of Bunsen and Kirchhoff is shown diagrammatically in Fig 13. The substance to be tested is heated at A, either by an open gas flame, or rather more conveniently, by electrical means. Light from the heated sample passes into the first tube of the instrument through a narrow slit at B, and thence through the collimating lens system C, which focuses it sharply at the prism, D. Here the light is split up in the way already described. It then passes into the second tube, where there is a further lens system E, the purpose of which is to spread out the spectrum band still more, and to emphasise the boundaries between the colours, which are normally rather diffuse. The image of the spectrum is focused at the eyepiece, F.

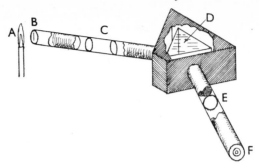

Fig 13   The essential parts of a simple spectroscope

If now the spectroscope is turned to a source of white light, such as an incandescent bulb, the familiar pattern of spectrum colours will be seen on looking through the eyepiece. But if the source of light is a gas flame into which a little common salt, or some other sodium compound, has been introduced, the spectrum band is seen to have disappeared. The field of view is dark, except for a brilliant yellow line (which a more powerful instrument would resolve into two lines close together). This line is in the region which would normally be occupied by the yellow part of the spectrum. If now some other element be substituted for the sodium, a different pattern of lines will be seen; this may be simple, like that of sodium, for some elements, but is extremely complicated, consisting of hundreds of lines, in different parts of the spectrum, for others—iron is a good example of this. The important point is, however, that for any element and a given instrument, this pattern will always be the same—it is, as it were, the 'fingerprint' of the element.

It is possible, though not particularly simple, to record this pattern by photography; a more practical method is to provide a scale in the eyepiece. This is generally graduated to represent wavelengths in Ångstrom units (see page 74), and it is thus a simple matter to record a spectrum for future reference. If then we introduce an unknown substance into the flame, and observe the lines which have been noted as belonging to, say, iron, chromium, and tungsten, we can assert that the sample consists of these elements; in other words, it is a particular kind of tool steel. In fact, since the elements must be vaporised to obtain a spectrum, the 'flame' in this instance would have to be an electric arc, which has a very high temperature, but the principle is unchanged.

Modern spectroscopes have many refinements, of course. A diffraction grating (see page 86), which also has the property of breaking up a beam of white light, is often used instead of a prism; it gives a better spectrum for many purposes. 'Straight-through' instruments, known as 'direct-vision' spectroscopes, are also made; they are more convenient for many purposes, and can be small enough to be carried in the pocket. Instruments of this kind are particularly useful for studying *absorption spectra*, and this important concept must now be explained.

As early as 1815, Joseph von Fraunhofer had shown that if sunlight is passed through a narrow slit, and then through a prism, the spectrum thus obtained is crossed by an enormous number of fine dark lines. The reason for this was a mystery until Kirchhoff explained it in 1859. He demonstrated that if white light, that is, light of mixed wavelengths, is passed through a gas, or the vapour of a heated liquid or solid, and then into a spectroscope, the spectrum will be crossed by dark lines, the positions of which exactly correspond to those of the bright lines which would be obtained by heating the element concerned. The explanation of this is fairly obvious; the unheated element *absorbs* those wavelengths which it would give out in its heated state. We shall not here go into the reasons for this, which were not properly understood for many years afterwards, but clearly, this provides an explanation of the Fraunhofer lines and at the same time gives us a research tool of enormous importance.

The sun consists of a very hot core, and a relatively cooler outer layer—though even the latter is very hot by terrestrial standards. Elements in this outer layer absorb light of the appropriate wavelengths from the light coming from the inner core, and since there are a great many elements, the number of the Fraunhofer lines is accounted for. By comparing the solar spectrum with the absorption spectra of known elements obtained in the laboratory, it is possible to show that all the elements known on earth also exist in the sun—in fact, one element—helium—was discovered in this way in the sun before it was shown to exist on earth. In the same way, by coupling the spectroscope to a telescope and pointing it at a star, we can find out what elements are present in the star, though it may be millions

131

of miles away—in fact, the nearest star, apart from the sun, is something of the order of twenty *billion* (2×10^{10}) miles distant. In making such analyses, allowance must naturally be made for substances present in the earth's atmosphere.

It is not essential that the elements to be examined by this method be in gaseous form. Liquids, and even transparent solids, give results, and we shall meet practical applications of this later. Both emission spectroscopy (with heated samples) and absorption spectroscopy can detect tiny amounts of an element. They are quick and simple in practice, and the identifications are very positive. Thus it is not surprising that spectroscopy finds many applications in science and industry, whenever we wish to know 'what is it made of?' It is not quite so well adapted to quantitative determinations, though an experienced spectroscopist can generally give an estimate of the amount of an element present from the relative intensities of the spectral lines. In this connection, the spectrograph is very useful.

This, in effect, is a mechanical spectroscope, in which the observer's eye is replaced by a photo-electric cell. This is caused to 'scan' the spectrum, and the lines are recorded as 'peaks' in the current coming from the cell. This current can be amplified and used to operate a pen recorder, producing a spectrogram, as shown in Fig 14. Interpreting this is a much more comfortable matter than peering through an eyepiece, of course; moreover, the spectrograph can do its work while the observer gets on with something else. It can even be arranged to analyse samples automatically at intervals, without any human intervention, and, by linking it to a computer, the results can be printed out in clear language without the necessity of any in-

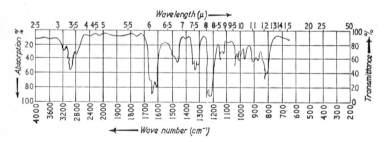

Fig 14   The infrared spectrogram of the substance methyl vinyl ether

132

terpretation at all—extremely useful in the control of certain industrial processes.

It is this principle which has made possible the tremendous advances in spectroscopy which have been made in the past few decades. Lack of space forbids our going into them in detail—not that they are 'difficult' once the basic principles are understood—but we will go a little way down the road. It is fairly easy to show that there is 'something' in the apparently dark regions beyond the ends of the visible spectrum produced by a prism. Thus, if a thermometer be moved slowly from the blue to the red end, it will record a gradual rise of temperature. If it is moved beyond the red, the temperature rises still further. Conversely, a photographic plate is affected more and more strongly as it is moved towards the blue and violet, and strongest of all in the dark region outside the violet. Clearly, these regions represent radiation akin to light, but which does not affect the human eye. The names *infrared* and *ultraviolet* have been given to these two regions respectively.

It is possible to build spectroscopes which operate with infrared and ultraviolet radiation, and for some purposes they are more useful than those working with visible light. Of course, it is no use peering through an eyepiece in an attempt to see these spectra. In principle, they could be made visible, but since detectors, very much on the lines of the photo-electric cell, are available, it is much simpler to produce the records as spectrograms.

5

Hot and Cold

The fact of heat was one of the earliest aspect of man's experience. It could scarcely be otherwise, for our bodies are equipped with sense organs specially for the detection of heat, and without them we could hardly survive, for the mammalian body can only work efficiently in a comparatively narrow range of temperatures. As a matter of fact, we have separate sense organs which respond independently to 'hot' and 'cold', and this fact may have contributed to the confusion which for long existed in the human mind both about the nature of heat, and the difference between heat and temperature.

We shall not go into the earlier theories and speculations about the nature of heat. We now know that it is a form of energy; to be more precise, it is a manifestation of the energy of motion. We saw in the last chapter that all matter is made up of tiny particles; atoms and molecules. The molecules of matter are in a state of constant movement, or vibration, smallest in solids, greater in liquids, and greatest of all in gases. Although it is not possible to see molecules, even with the most powerful of microscopes, the effect of this vibration can be observed, and its existence thereby proved experimentally. It can also be demonstrated that as a substance is heated, the movement of the molecules increases. This increase of movement, which is heat, also takes place if the substance is rubbed hard, or if an electric current is passed through it, and in fact, we can say, with but little violence to the real facts of the situation, that all energy tends eventually to manifest itself as heat.

That heat and temperature are not the same thing is easily shown. Suppose an iron bar is heated for a definite length of time, and then plunged into, say, a gallon of water. The water will be heated to a certain degree. If, now, the same bar is heated for the same length of time, but this time is plunged into two gallons of water, the water will only become half as hot. The same *amount* of heat has been put into it, but since there was more water, the *temperature* is lower.

Although we can sense heat and cold, our equipment for the purpose is relatively crude. The human body is, as a matter of fact, equipped with an extremely accurate temperature-sensing device, but it is part of our general temperature-regulating mechanism, which is entirely automatic, and outside conscious control. This mechanism maintains our body temperature constant to within very fine limits; better than could be managed by most man-made devices, and this very perfection means that our awareness of heat and cold need only be of the most approximate kind, sufficient simply to keep us away from extremes with which the body's temperature controls could not cope. Moreover, our experience is largely of a comparative nature, as can easily be demonstrated by the following well-known experiment. Place one hand in hot water, and the other in cold. After a minute or so, place them both in lukewarm water. This will feel cold to the hand which was in the hot water, and hot to the one which was in the cold!

Subjective judgements of temperature, that is, the *degree* of hotness (coldness can now be disregarded as a purely human experience which has no real counterpart in the outside world) are thus not very reliable. This did not matter much until the rise of science. In the arts a method of accurate temperature measurement would have been useful, but the necessity was not such as to drive anyone to invent it. Temperature control in technical activities was mainly needed in high-temperature processes, such as the working of metal and glass, and at these temperatures the experienced workman can make quite accurate judgements from the colour of the heated material.

Most metals are oxidised if they are heated, that is to say, they combine chemically with the oxygen of the air. If a piece of steel, say, is made bright and then heated, a thin film of oxide will form upon its surface, and the colour of this film changes as the metal is heated, making it possible to judge the temperature fairly accurately. It will first of all be a pale straw colour; this gradually darkens to brown, and then begins to be suffused with purple. As the temperature is increased, the purple covers the whole piece, and then turns to blue:

first dark and then light; finally it becomes white. These changes are not reversible; if the heating is stopped at any point by suddenly cooling the piece of steel in water or oil, the colour remains, a fact that is utilised in many branches of engineering. However, if heating is continued, a reversible change sets in: the metal begins to glow. At first it glows dull red; this brightens to a cherry colour, which passes into bright red, then successively to orange, yellow, and white; the lighter the colour the higher the temperature in each case. Heating beyond this point will cause steel to melt, and some other metals will have melted well before white heat is reached. This fact gives another way of judging temperature: long before the matter was proved, it was realised intuitively that certain substances always melt or boil at the same temperature. These methods served the purposes of early technology, and are indeed far from obsolete today.

All substances expand when heated, the expansion being in general greatest in gases and least in solids; it varies from material to material. The recognition of this fact pointed the way to the construction of an instrument for measuring temperature. This invention has been ascribed to a good many people, but it seems that Galileo has the best claim. The date of the invention is uncertain; many of Galileo's papers were destroyed by the church, with which he came into conflict, and it is thought that his records of this work were amongst them. A date of 1592 has been suggested, and is probably not far out.

Galileo's *thermometer* used the principle of the expansion of a gas —air. It consisted of a length of glass tube, with a closed bulb formed at one end of it. The other end dipped into a vessel containing water. The bulb was warmed, for example by holding it in the hand, and the air inside expanded, bubbling out of the open end of the tube. When the source of heat was removed, the air contracted, and the water rose in the tube, to a distance proportional to the temperature to which the bulb had been heated.

The use of the hand as a heating medium suggested one of the earliest practical uses of the thermometer. Doctors had long realised that the body's temperature is raised in fever, and they were not slow to make use of the new instrument, or variations of it, for measuring body temperature. The tube had to be graduated for this purpose, but the graduations were at first quite arbitrary. Because of the great *coefficient of expansion* of most gases, gas thermometers, of which the air thermometer is an example, can measure very small changes in temperature, but in the form just described, they have one most serious defect, which is that their indications are dependent not on temperature only, but also on atmospheric pressure (barometric pressure) which varies considerably. This may well have been known to Galileo himself, for he did work on atmospheric pressure, but in any case it was soon recognised, and attempts were made to construct a thermometer which employed a liquid instead of a gas, and a sealed tube to isolate it from the air. Such instruments were being made in Florence by 1670. The essential features are unchanged today: there is a glass tube with a bulb at one end, in which a quantity of liquid is sealed. The tube above the liquid is evacuated. As the bulb is heated the liquid rises, and as it is cooled it falls. The tube is provided with a suitable scale, of course.

Water is not a suitable liquid for use in a thermometer; not only does it freeze at a temperature frequently encountered in many countries, but it also does not expand and contract evenly. If water is cooled, it contracts, just like any other liquid, until it is a little above the freezing point, but then it begins to expand again until it freezes, which fact is responsible for many burst pipes. The early workers tried to use mercury in their sealed thermometers, but abandoned it because it did not expand sufficiently. They finally settled on alcohol, which is satisfactory for meteorological purposes, since it does not freeze until well below the range of temperatures reached in most countries, and which has a good expansion. Most cheap room thermometers in use today contain alcohol. It is not much use for high temperatures, however, since its boiling point is too low.

All that was now needed was a system of graduations which could be reproduced anywhere. The idea of 'fixed points' seems to have been suggested by Robert Boyle, and the freezing and boiling points of water were used as such points by Huygens; this was around 1660–70. Gabriel Daniel Fahrenheit, who was born in Danzig, also devised a thermometric scale, in connection with a very accurate alcohol thermometer which he constructed, some time before 1709. He also perfected the design of the mercury thermometer, first by using much finer bore tubing, which overcame the difficulty of the relatively small expansion of mercury, and second, by devising a method of purifying mercury. This instrument is still the standard in many laboratories all over the world.

Fahrenheit's work on the thermometric scale was less happy. For the zero of his scale, he took the temperature of a mixture of ice and salt, which was the coldest thing he knew, and which he erroneously supposed to be the coldest attainable. There were two other fixed points: the freezing point of water and the temperature of a healthy human body. The latter is now known to vary slightly, in a daily cycle, so it was not a very good choice. The distance between the upper and lower points was originally divided into 96 degrees; this made the freezing point of water 32 degrees. Later the scale was slightly amended to give 180 degrees between the freezing and boiling points of water, which were recognised to be better fixed points, so the boiling point of water is 212 degrees, and the temperature of the body is generally taken as 98·4 degrees.

The shortcomings of this scale were recognised by Anders Celsius (1701–44), a Swedish astronomer, who suggested a scale based upon the definite fixed points of the temperature of melting ice and that of water boiling under known conditions of pressure. The scale was to be divided into 100 degrees, with the boiling point as zero and the freezing point as 100. This arrangement was reversed by a Frenchman, Jean Pierre Christin, to give the centigrade, or Celsius, scale as we have it today. A third scale, the Réaumur, had 80 degrees, but it is now obsolete, and will not be further considered in this book.

We thus have two scales, one complicated and rather irrational, the other simple and rational. Needless to say, the former was seized upon with great glee by the English, who are only now, slowly and with much difficulty, being persuaded to abandon it. The rest of the world, practically speaking, adopted the centigrade scale, which is also invariably used for scientific purposes. As far as possible, temperatures in this book are given in centigrade degrees—°C, meaning 'degrees Celsius'.

TYPES OF THERMOMETER

The ordinary room thermometer is familiar to everyone. The working fluid in such an instrument is coloured alcohol; because of its relatively large expansion, the glass tube can be of a fairly large bore, making the liquid easy to see. The scale is printed on the wooden mount, so the accuracy of the indications obviously depends, in the first place, upon the positioning of the tube on the mount. Normally, it is good enough for the purpose required, but for more accurate work in meteorology, an instrument more on the lines of the laboratory thermometer is used.

A laboratory thermometer is shown in Fig 15. The working fluid here is mercury, so the bore of the tube is finer; both the tube and the bulb are also much more accurately made. This instrument must be capable of being immersed in liquids, so the scale is actually etched on the glass tube itself. Moreover, a good instrument is individually *calibrated*, that is, the upper and lower fixed points are marked by reference to the freezing and boiling points of water. Thermometers of this kind are made in a variety of patterns, to cover various ranges of temperatures, and to read to greater or smaller degrees of accuracy. It is not only in the laboratory that such instruments are needed: some kinds of photographic processing require that solutions shall be maintained within a fraction of a degree of the specified temperature, and 'certified' darkroom thermometers, guaranteed to within 0·1° C, are obtainable.

It is not always convenient—or conducive to accuracy, human limitations being what they are—to observe a thermometer constantly. As with many other instruments, a method of recording, or at least of retaining, a given reading, is desirable, and the next two instruments which we shall consider do this in different ways.

Figure 16 shows a maximum and minimum thermometer, used for meteorological purposes, and also in greenhouses. The working fluid in this instrument is alcohol, though the tube is also seen to contain mercury. The function of the mercury in this case, however, is simply to transmit motion; in a wide-bore tube its own expansion can be neglected by comparison with that of the alcohol. As the temperature rises, then, the alcohol in the bulb at the left expands, and in doing so it pushes the mercury down in the left-hand limb of the U-tube, causing it to rise in the right-hand limb. Above the mercury is a tiny steel *index*, shown greatly enlarged in the diagram. As the mercury rises, therefore, it pushes the index up the right-hand tube,

139

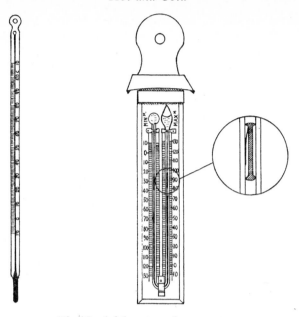

Fig 15   A laboratory thermometer
Fig 16   A maximum and minimum thermometer with (*inset*) an enlarged
view of one of the indices

steel being lighter than mercury, but when the temperature falls, and the mercury returns, the index stays at the highest point which it has reached, being held by the tiny spring pressing on the wall of the tube. The alcohol in the right-hand tube serves merely to balance the system; it can flow freely past the index. Thus we can inspect the thermometer at some future time, and read from the position of the index the highest temperature that has been reached. In the same way, as the alcohol contracts with falling temperature, the mercury returns, impelled by the pressure in the right-hand tube, and the position of the left-hand index gives an indication of the lowest temperature reached. The indices are reset by means of a magnet after the maximum and minimum temperatures have been read and recorded.

As has already been mentioned, recording the temperature of the human body was one of the earliest practical uses of the thermometer, and nowadays the checking of temperature is one of the most important routine items in medical diagnosis and treatment, as anyone who has been in hospital knows quite well. It is inconvenient, to say the least, to try to read the patient's temperature from a thermometer placed somewhere in or on his body, and a clinical thermometer is arranged so that it will retain its reading after it is removed from the source of heat. Its general construction is shown in Fig 17. Everyone should know how to read a clinical thermometer, and while the method is best learnt from a demonstration by a doctor or nurse, the following will serve both as a description of the process and an explanation of the working of the instrument.

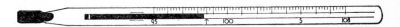

Fig 17 A clinical thermometer for Fahrenheit readings. There is, of course, a connection between the mercury in the bulb and that in the stem, but it is normally invisible to the naked eye, the bore being narrowed to form the constriction, as described in the text

As will be seen from the illustration, this type of thermometer is intended to record temperatures in a comparatively narrow range on either side of the 'normal' body temperature only. This 'normal' temperature of about 37° C (98·4° F) is usually marked by a special graduation on the scale, but it should not be taken as being anything more than a rough indication of normality. In order to get this restricted range, together with a suitably 'open' scale, so that very small differences in temperature can be read, the bore of the tube is very fine. To make it easier to read, the outside of the tube, opposite the scale, is formed into the shape of a lens, which serves to magnify the width of the very slender thread of mercury.

The thermometer having been sterilised by immersion in some suitable antiseptic (not in boiling water!), the bulb is placed, as a rule, in the patient's mouth, though it is sometimes better to put it under the armpit, especially with children. If in the mouth, it must go under the tongue, and the lips should be closed. It is allowed to remain in position for a sufficient time—about a couple of minutes. Some thermometers are marked '½ min', but it is usually reckoned that this time is hardly long enough for a true reading. The instrument is then removed to be read.

Now, an ordinary thermometer, under these circumstances, would lose heat and the mercury would immediately begin to fall. This is prevented in the case of the clinical thermometer by the incorporation of a *constriction*, which is a narrowing of the tube just above the bulb. The mercury can flow past the constriction when it is forced along the tube by expansion, but immediately the bulb cools slightly and the mercury in it begins to contract, the thread of mercury breaks at the constriction, and the mercury already in the tube remains there, thus retaining the reading. Just below the lowest graduation, two parallel lines will be seen engraved on the tube. The instrument is held in a good light—though not over a candle, as in a certain well-worn story—and slowly rotated, until, at a certain point, when the lens-shaped part of the tube is in the correct position, the mercury column suddenly becomes visible between the lines, when the temperature can be read from the scale.

This done the mercury must be returned to the bulb, and this is done by shaking the instrument sharply. There is a decided 'knack' in this, and it is best learnt by a demonstration. When it is correctly done, however, the mercury will flow back past the constriction, and the instrument is ready for use once more. Larger instruments, working on exactly the same principle, are used in other situations where it would be awkward to read a thermometer in position.

HIGH TEMPERATURES

The boiling point of mercury is 357° C. This is not much above the melting point of lead, and well below the melting points of many common metals. Since no other convenient liquid with a higher boiling point is available we have to turn to other methods for the measurement of high temperatures. Instruments for measuring high temperatures, which as a rough guide may be defined as anything over 400° C, are usually called *pyrometers*. The early ones made use of the principle of expansion, just as do thermometers, but used a solid instead of a liquid. An illustration of such a pyrometer, such as might have been used for measuring the temperature of a furnace, is shown in Fig 18. The principle here is quite simple: the working element is the iron rod contained in the porcelain tube on the left. This is inside the furnace; a porcelain rod passing through the furnace wall transmits the expansion of the metal rod to the outside, where it is magnified by the gearing, and can be read on the scale.

Obviously, this arrangement cannot be used above the melting point of the metal used for the element, and though there are metals with very high melting points, they are correspondingly difficult to extract from their ores, and are consequently too expensive to be used in pyrometers. Firebrick, carbon, and similar *refractory* substances do not melt even at very high temperatures, but their co-efficients of expansion are too low to be convenient for temperature measurement. Other means have therefore had to be devised.

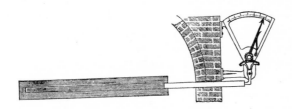

Fig 18   An early pyrometer

The first of the methods which will be discussed here depends upon a phenomenon called the *thermo-electric effect*. If a junction is made between wires of dissimilar metals, as shown in Fig 19, the outer ends of which are connected to a sensitive meter, a current will be seen to flow when the junction is heated. The size of this current is a function of the temperature and the metals composing the junction, which is generally referred to as a *thermocouple*. Strictly speaking, it depends on the *difference* in temperature between the junction and another part of the circuit, but for measuring high temperatures it is sufficient to keep the meter at room temperature and to assume that this is constant. There is a wide choice of metal

Dissimilar metals

Fig 19   The principle of the thermocouple

combinations for the thermocouple, and a suitable pair is selected for the expected temperature range and the degree of sensitivity required. One very popular combination is iron and constantan, which is an alloy of nickel and copper. This thermocouple can work up to 800° C, and gives a fairly high voltage, that is, it is quite sensitive. It must be realised that a 'high' voltage in this connection is not great by ordinary standards; the voltages from thermocouples are measured in millivolts, so sensitive meters are required.

Thermocouples are not necessarily confined to use at very high temperatures; provided the difference in temperature between the thermocouple and another part of the circuit can be arranged, and a suitable pair of metals is chosen, temperatures can be measured within practically any desired range. The beauty of this arrangement is that the recording meter does not necessarily have to be anywhere near the actual thermocouple. The advantage of this at high temperatures needs no stressing, but it is often very useful in other circumstances as well.

Another useful device which enables us to measure the temperature of something (a furnace, for example) at a very high temperature without getting too near it is the optical pyrometer. This is an ingenious piece of equipment based on the phenomenon of colour change with temperature, described earlier. The essential part is a thin piece of wire, generally arranged in zigzag form. This wire can be heated by passing an electric current through it. The heavier the current, the hotter the wire becomes, and the brighter it glows, going through the usual stages of dull red, bright red, yellow, etc, to white. The temperature and the brightness are both related to the current, and the latter is shown by a meter, or the instrument can be calibrated in advance, and the temperature of the wire shown on a dial attached to the control knob. The wire is arranged so that it can be viewed through an eyepiece. If now the instrument is pointed at a black surface, and the current is switched on, the wire stands out brightly when viewed through the eyepiece. In the same way, it will appear dark if viewed against an open furnace, for example. Now, if the current is gradually increased, the wire gets hotter and brighter, until it is just as bright as the furnace, at which point it can no longer be seen. If the current, and thus the temperature, is raised still further, the wire again becomes visible, but now it appears brighter than the furnace. So, all we have to do to measure the furnace temperature from a distance is to adjust the control so that the wire just disappears, meaning that it is at the same temperature as the furnace, and then read off the temperature on the dial.

So much for measuring high temperatures. But how do we attain such temperatures, and what is the point of doing so, anyway? Well, one of the reasons for heating something is just to see what happens. This is not quite as silly as it sounds. Men have been doing it probably since the discovery of fire; certainly since the discovery of cooking, and a vast number of useful discoveries have been made in this way. Apart from that, it is known that some of the fundamental processes of the universe take place at very high temperatures, in the interiors of the stars, and in order fully to understand these processes, it will be necessary to duplicate them in the laboratory. As yet, we are far from being able to do this, but high-temperature research is a necessary stepping-stone to it.

For several thousand years, the burning of fuel, which in practice means some form of carbon, provided virtually the only way of reaching high temperatures. The faster the burning, or *combustion*, the higher the temperature, generally speaking. The temperature of our bodies is maintained by a very slow combustion of the carbon in certain foods; the oxygen, which is necessary for all combustion, coming from the air. The carbon dioxide which is present in exhaled breath, as mentioned on page 116, is a product of this process. When combustion is faster, heat is produced at a greater rate, temperature rises, and light as well as heat is produced; in fact, we have a fire. The temperature which can be attained in this way depends, broadly speaking, on two things: the *calorific value* of the fuel, and the rate at which it can be made to combine with oxygen. A detailed explanation of the concept of calorific value would take us into the realms of calorimetry and thermochemistry, and will not be attempted here. Briefly, and rather loosely, it means the amount of heat which can be got from a definite quantity of the fuel when it is completely burnt.

The rate of combination with oxygen depends primarily upon the rate at which oxygen can be supplied. The air above an open fire becomes heated, and rises, hot air being less dense than cold, and fresh air—and therefore oxygen—flows in to replace it. A large fire will therefore be hotter than a small one, other things being equal, since this process proceeds faster. However, the highest temperature likely to be attained in an open fire is about 700° C.

Everyone has noticed how an open bonfire burns more vigorously when it is fanned by the wind. This fact was used in the building of furnaces at a very early date: it was the custom to site them on hill-sides, where the prevailing wind could provide the draught. Furnaces of this kind were used to smelt copper, and sometimes even iron, which requires a temperature of over 1,000° C. Suitable sites for this kind of furnace are not plentiful, and the force and direction of the wind are very variable quantities, so a more controllable means of providing extra draught was sought. Furnaces blown by bellows were a very early invention, and for many hundreds of years the tempera-tures of 1,200° C or thereabouts obtained by this means were the highest which could be artificially reached.

Though it is by far the commonest, carbon, generally in the forms of wood, charcoal, coal, or coke, is not the only possible fuel. During the nineteenth century it was realised that much more efficient com-bustion, and higher temperatures, were possible if the fuel was in the form of a gas. Most fuel gases still contain carbon, however. The open-hearth furnace, used in steelmaking, uses gaseous fuel, and the fuel and air are preheated before being burned. A temperature of 1,600° C can be reached in a furnace of this type, and this is about the limit using a fuel/air mixture. For higher temperatures we must use pure oxygen, preferably in conjunction with a fuel gas. Hydrogen was the first one to be tried, and when burned with pure oxygen in a special *blowpipe*, it gives a flame temperature of 2,200° C, which will melt steel quite easily. But the combination giving the highest practical temperature is oxygen and acetylene. Acetylene is a gas containing carbon and hydrogen, and the temperature of an oxy-acetylene flame is 3,100° C. This, for practical purposes, is the highest temperature attainable by a combustion process. The oxy-acetylene blowpipe is extensively used for joining metals by *welding*; the parts are literally fused together.

To attain higher temperatures we have to turn to other methods. In 1802 Sir Humphry Davy discovered that if a piece of hard carbon is connected to each pole of an electric battery, and the pieces are then touched together and rapidly separated again, a brilliant flame, or spark, leaps continuously across the gap. This is the electric arc; it was known to be extremely hot as well as a source of light, but its practical exploitation had to wait upon the development of a more efficient source of current than could be provided by batteries. Once this was available, use of the electric arc as a source of heat became possible, and nowadays it is extensively used, in welding, in furnaces, and for many other purposes. The temperature of an arc is about 3,500° C.

The highest temperatures so far achieved in the laboratory are in excess of a million degrees Celsius. They are reached by means of what is called a *pinched plasma discharge*. A very large electric charge is built up in a bank of *capacitors*, which are devices for storing electricity, and the entire charge is released in a few microseconds, giving a current of the order of a quarter of a million amperes. This current is discharged between electrodes in a gas at low pressure—deuterium, which is an isotope of hydrogen, is usually used—and raises the temperature of the gas to the point mentioned. At this temperature substances can no longer exist as normal gases; the atoms are virtually broken apart, and the resulting mixture of electrons and nuclear particles is called a *plasma*, sometimes referred to as 'the fourth state of matter'—the three normal ones being solid, liquid, and gas. Plasmas can respond to the magnetic effects of an electric current, and in this case the result is that the plasma is forced into a narrow thread between the electrodes, away from the walls of the container. This is the 'pinch effect'. It is a fortunate phenomenon, for otherwise the hot plasma would be rapidly cooled by the container walls, or, if it could be maintained long enough—so far, it cannot—would itself vaporise the container. For the same reason, the temperature cannot possibly be measured directly; it has to be deduced from other phenomena produced.

At present, this is only a research experiment. Its significance is, that if we could obtain and maintain a sufficiently high temperature under controlled conditions, power from nuclear fusion—the basis of the hydrogen bomb—would be a real possibility. This would give us a source of power which, by human standards, would be virtually inexhaustible—it is the source of the radiation of the sun and other stars.

Those last words offer a clue to a further method of obtaining high temperatures. The sun has an 'atmosphere', called the *corona*, which is a plasma of the kind described above. Its temperature is about a million degrees. The actual surface of the sun is at a temperature of 6,000° C; the temperature of the interior must be very much higher, but can only be guessed at. We cannot, of course, bring these temperatures directly to earth, yet the sun is pouring out radiation at an unimaginable rate. That tiny fraction of it which reaches the earth at a distance of (in round figures) 93 million miles, suffices to maintain all life on this planet. The idea of making use of the sun's heat for technical purposes is an old one: there is a story, probably apocryphal, that Archimedes, who lived around 287–212 BC, and who was one of the few Greek philosophers to take an interest in experimental science, used giant mirrors to set fire to a Roman fleet at Syracuse. This seems rather unlikely, since the technology of the time would probably not have been equal to making such things, but it is entirely possible that he knew that the sun's heat could be concentrated in this way.

The knowledge seems to have been lost or disregarded until the invention of lenses, when it was realised that a convex lens could also be used to concentrate the sun's heat. There can be few of us who have not played with a 'burning glass' in our schooldays. The larger the glass, the more heat it will receive, of course, and this, concentrated into a small spot, can yield very high temperatures. The English chemist Joseph Priestley (1733–1804), who discovered oxygen, made many experiments with a very large burning glass which a patron had given him. This had a diameter of 6in, a technical triumph for the year 1774, when these experiments were reported. However, even today, a large lens is by no means a cheap or an easy thing to make, and with the invention of gas burners, burning glasses passed into disuse as laboratory equipment. The idea of making practical use of the sun's heat was never quite forgotten, however, and various schemes were devised from time to time, and some were even tried out. Thus, there was a fairly successful solar boiler installation in the Sahara before the Second World War. None of these schemes came to much, however, and as in any case they were not concerned with the attainment of very high temperatures, we can pass them by in this book.

As mentioned above, a very large lens is not an easy thing to make;

Page 149 A commercial helium liquefier, the Philips PLHe–209. This is a fully automatic unit which can produce 8 to 9 litres of liquid helium per hour. The essential refrigerating elements are the two 'cryogenerators' on the left and to the rear; the liquid helium is delivered to and stored in a Dewar vessel, the top of which can just be seen on the lower right

Page 150 Shrink fitting cylinder liners into motor engine cylinder blocks at the DAF works. The liners are cooled in liquid gas in the cabinet on the left. Apart from its convenience, this method has the advantage that the highly-finished liners are not distorted or discoloured, as they might be if the blocks were heated

a lens of 40in diameter for an astronomical telescope is regarded as representing the practical limit, and larger telescopes employ mirrors, the largest of which so far made is of 200in diameter. Of course, it has always been known that a concave mirror could concentrate the sun's heat in the same way as a burning glass, but it does not seem to have been realised until quite recently that here was a means of reaching high temperatures quite easily and cheaply. A mirror for this purpose need not be nearly so accurately made as one for a telescope, and it can be made of metal instead of glass, so the limit of size is set, for practical purposes, only by the problems of supporting and aiming it. Several of these 'solar furnaces' have been built in recent years. They can of course only be fully utilised in tropical, and preferably in desert, country, where long hours of continuous sunshine can be relied upon.

The detailed design of these devices differs, but basically they all consist of a large concave (strictly a parabolic) mirror, with mechanism for enabling it to follow the apparent movement of the sun, and means of holding objects to be heated at the focus. One such furnace at Bouzareah, near Algiers, has a mirror 8·4m in diameter, and can concentrate 40kW of heat, at a temperature of 3,000° C on to a reaction vessel at the centre, and can keep on doing it for hours on end, for the expenditure of no more power than is needed to keep it pointing in the right direction. Another very interesting feature of this furnace is the facility to select only a part of the solar spectrum, so that, for instance, the effects of very intense radiation with ultraviolet light at high temperatures can be studied. It is used mainly for chemical research.

Some of the reasons for wanting to attain high temperatures have already been mentioned; most of them were concerned with fundamental research. There are, however, a good many more practical applications but only a few can be sketched here. Modern technology is itself tending to make use of higher and higher temperatures, and it is necessary to find both the means of attaining them and measuring them, and materials which will stand up to them. *Refractory* materials have, in the past, generally been some form of ceramics, but these, useful though they still are, have drawbacks for some purposes. Much high-temperature research is devoted to finding new materials, and in particular, metals and alloys, which will withstand the temperatures met with in space vehicles, both for rocket motors and to deal with the very high temperatures generated in 're-entry' to the atmosphere. Some notable successes have already been achieved.

One of the earliest applications of high-temperature technology was the manufacture of artificial gemstones. Sapphires and rubies came first. These stones are both crystallised alumina, aluminium oxide, the difference in colour being due to very slight traces of different impurities: chromium in the ruby and titanium in the sapphire. Soon after the invention of the oxy-hydrogen blowpipe (page 146), it was found that it could be used to melt ruby and sapphire, and attempts were made to manufacture usable gemstones by fusing together dust and chippings, the waste products of the cutting of natural gems. These experiments were fairly successful; in fact some of the stones so made are still encountered by jewellers today, but it was not long before a method was devised of making synthetic rubies (these are more in demand than sapphires) from the basic raw materials.

This process is still operated today. The stones are made in a small furnace, the heat of which is provided by an oxy-hydrogen blowpipe which burns continuously for several days. At the top of the furnace, powdered alumina, mixed with a carefully calculated quantity of chromium oxide, is contained in a small sieve. This is rapped by a mechanically-operated hammer at intervals, causing a small quantity of the powder to fall through the flame, where it melts and deposits on a pipeclay stem placed to receive it. This stem is withdrawn from the flame very slowly, also over a period of days, and as it is withdrawn, the artificial ruby *boule* builds up on it. Rubies made thus are not exactly cheap, because of the cost of the gases, and the time taken by the process, which must be carefully controlled all the time, but they are cheaper than the natural product, with which they are identical chemically. An expert, however, can usually tell the difference, and more will be said about this in Chapter 9.

High-quality emeralds are even more valuable than rubies. Emerald is a form of the mineral beryl, a silicate of beryllium and aluminium, and synthetic emeralds have been made. It is certain that high temperatures must be involved in the process, but the exact method is a closely-guarded secret. As a matter of fact, the principal object nowadays of making these substances artificially is not their gem qualities but their hardness. Ruby and sapphire are both forms of corundum, an impure form of which is emery. Both are used extensively to make bearings in watches, where extreme wear resistance is important. Emery is not produced artificially, however, for a completely synthetic substance, carborundum, or silicon carbide, is even harder. This is a product of the electric arc furnace, and is made in great quantities as an abrasive.

In view of all this it is natural that the attempt should have been made to manufacture the hardest substance of all—diamond. Diamond is crystallised carbon, the melting point of which is over 3,500° C, so it is not surprising that most attempts to produce diamond in the laboratory have failed. Recently, however, a commercially successful process, based on the electric furnace, has been developed by the General Electric Co in the USA. Full details have not been released, but it is fairly certain that there is more to the matter than high temperature—high pressure is also involved. Diamonds made by this process are microscopic—like dust to ordinary appearances—but they are undoubtedly diamond, and are finding increasing use in industry, where diamond dust is a valuable abrasive and polishing medium. As yet, however, no one has succeeded in making diamonds of gem size and quality.

CRYOGENICS

The word above may be a new one to many readers, and it is in fact quite a newcomer to the scientific vocabulary. It is a useful 'omnibus' expression, meaning the science and technology of reaching and using very low temperatures, so we are now turning to the other extreme of the temperature scale. However, we come at once to a difference more profound than that of temperature. While there is, so far as we can tell, no limit to the highest temperature that can be attained, there is a definite limit to the lowest.

We saw on page 134 that heat represents the movement of the molecules of matter, and that as temperature is increased, this movement speeds up. Conversely, as the temperature is reduced, the movement slows down, until finally, at a sufficiently low temperature, it would stop altogether. At this stage there would obviously be no more heat, and a lower temperature than this is therefore a meaningless concept. This lowest attainable temperature is called *absolute zero,* and if it were to be attained, it would be —273° C. That 'if' is a very significant one; it can in fact be shown theoretically that, while absolute zero can be approached ever more closely, it can never quite be reached. The human mind finds it quite difficult to accept a situation like this, but there are many examples of such limitations, and to that extent the human mind is out of tune with the real universe. What is really meant when we say that something is 'impossible' in science is that if it were to happen, the normal laws of physics, as we understand them at present, would no longer be valid. This, however, is a matter for the philosopher rather than the scientist. Absolute zero has never in fact been attained, although, as we shall see, it has been approached very closely indeed.

However, since we can calculate, if we cannot reach, the temperature of absolute zero, we have the starting point for a completely rational temperature scale. It is logical to call absolute zero 0°, since there can be no temperature lower than this. As there is no corresponding upper limit, the size of the degrees must be chosen more or less arbitrarily, and the degrees of the centigrade scale are used, as scientists are already accustomed to working with them. The scale so constructed is called the Kelvin scale, in honour of Lord Kelvin (1821–1907), a British scientist who did important work in thermodynamics, and temperatures on it are expressed as °K. This scale is particularly useful in low-temperature work, as it does away with the necessity of using minus signs and 'counting backwards', but it is also extensively used in other branches of physics. Zero on the Celsius scale is 273° K, and 'room temperature' is conventionally taken as 300° K, which makes for easy calculations.

The attempt to reach lower and lower temperatures has been going on for rather more than a hundred years. There is no space here to go into the history of this research, but the story is well told in Dr Mendelssohn's fascinating book (see Reading List). Suffice it to say that the starting point, and for a long time the principal reason for the quest, was the desire to liquefy gases. The earliest attempts to do this consisted in subjecting the gases to greater and greater pressures, but while some gases were liquefied in this way, the majority could not be, and it was eventually realised that temperature, as well as

154

pressure, was an important factor. In fact, for any given gas there is a definite temperature above which it cannot possibly be liquefied by pressure alone. This is the *critical point* of the gas in question. For carbon dioxide it is 31° C, for ammonia, 130° C, for chlorine, 141° C, and for sulphur dioxide 155° C. These temperatures are all well above room temperature, which explains why these gases were liquefied quite early by the application of pressure only. However, the critical temperature of oxygen is 155° K, or −118° C, and in order to liquefy it, it must be cooled down below that temperature. How can this be done?

To answer this question we look first not at a very low-temperature, but at a comparatively high-temperature phenomenon. Most people know, through experience in pumping up bicycle tires and such-like, that when a gas is compressed it gets hot. The reverse process is not so immediately obvious, but it is a fact that, for example, the exhaust from a steam engine is at a considerably lower temperature than the boiler steam, as well as being at a lower pressure. The steam has done work in the engine, and in the process it has expanded and at the same time lost heat. This also is a general rule: when a gas expands, it gives up heat.

Now, when we want to raise the temperature of something, we put heat into it from a flame, an electric heater, or some other heat source. To lower the temperature, we have to extract heat, and the processes described above give us a means of doing so. Suppose we have a compressor, compressing some gas, which could very well be ordinary air. The compressed gas will be hot, so we arrange for the discharge pipe from the compressor to pass through a cooling water bath. Here heat is extracted from the gas, and by having a constant flow of water, the gas is cooled to the temperature of the water.

The compressed gas is then passed on to an expansion engine, which is in effect a compressor in reverse, in other words, it very much resembles a steam engine. Here the gas does work against the piston of the engine, and in so doing it expands and loses heat. The discharged gas is therefore cooler than the gas which enters the engine, but instead of exhausting it to the atmosphere, as in the case of a steam engine or compressed air motor, it is taken back to the compressor for the cycle to be repeated all over again. On the way back, the pipe carrying the cooled gas passes round the one bringing the gas into the engine. This arrangement is termed a *heat exchanger*, and the effect of it is to ensure that the gas coming into the engine for the next stroke is cooler than for the previous one. Thus the whole process is cumulative, and eventually the gas becomes so cold that it liquefies in the engine.

When it is first encountered, this process appears to savour of lifting oneself by one's own bootstraps, but in fact it is not so: what we are doing is to move heat from one part of the circuit (the engine) to another part (the cooling water) where it is discarded, power being applied to the compressor to bring about the movement. The process is perfectly logical, and easy to understand if the difference between heat and temperature is kept clearly in mind.

This is a very efficient device for producing low temperatures, but a piston engine is mechanically complicated, and the engineering problems are aggravated by the behaviour of materials at low temperatures, as we shall shortly see. In large steam-driven installations, such as power stations and ships' engines, the piston engine has nowadays been supplanted by the turbine, which is a purely rotary device, and much more efficient. The same change has taken place in large aircraft. It seems logical to use a turbine instead of a piston engine in the liquefaction apparatus, and this is in fact done in many cases. There is another way round the problem, however. The essential step in bringing about the cooling effect is the expansion of the gas. This has been envisaged as taking place against a piston. But this is not an essential; if a compressed gas is free to expand, it will do so, whether or not the piston is present. So we can dispense with the engine altogether, and simply allow the gas to expand through a nozzle. This is a much simpler arrangement mechanically, but it is also less efficient, and many more cycles are required to bring about liquefaction. For many purposes, however, the price is one which can be paid.

Before looking at the practical uses which are made of these principles, it will be as well if we consider at this stage the ordinary refrigerator. Both the familiar domestic machine and the large installations used in cold stores, ice-cream factories, and similar places work on the same principles, which are similar to, but not quite identical with, those already described. There are two principal types, compression and absorption machines. Only the compression type will be described here, owing to limitations of space. The heart of such a machine is the compressor. In this is a suitable gas, which may be ammonia, sulphur dioxide, or, as is usual in domestic machines, dichlorodifluomethane, usually, and much more conveniently, known by the trade name of Freon. The gas when compressed gets hot, as usual, but instead of water cooling, it is usually cooled by passing it through a long finned or convoluted pipe, so that heat is lost to the surrounding air, which is simpler.

The gases chosen for refrigeration have critical points well above normal room temperature, so the gas, now being cooled to room temperature and under pressure, condenses to a liquid. The liquid passes through a valve, which reduces the pressure, into a coil of pipe surrounding the cold chamber of the refrigerator (the freezing compartment in a domestic machine). Since it is under a reduced pressure it turns back into a gas, or in other words, it boils, the heat needed for this process being extracted from the cold chamber, the temperature of which falls in consequence. The gas then passes back to the compressor and the cycle begins all over again. Left to itself, the process would continue until the cold chamber had not sufficient heat left to boil the liquid, when, all the gas being turned into liquid, and liquids being incompressible, the compressor would stall. This is undesirable, and in any case it is not usually necessary to make things as cold as this, so a thermostat is incorporated, which switches off the compressor when a predetermined temperature has been reached. This also economises the driving power, which is generally electricity.

Machines such as these can reach temperatures well below the freezing point of water, which is all that is needed for food preservation and similar purposes. For gas liquefaction, however, it is necessary to get down to temperatures much lower than this, so machines working on the principles first described are used. But why liquefy gases anyway, it may be asked. One reason is simply as a means to an end. Liquid gases offer one of the most convenient means of studying low-temperature phenomena, and also offer a way of reaching still lower temperatures. Thus, suppose we have a machine for liquefying carbon dioxide. As this has a critical temperature of 31° C, such a machine is easy to build and operate. Now, if the liquid carbon dioxide so obtained is used in place of the cooling water in a second machine, it is obvious that this machine will be able to reach much lower temperatures. This process is referred to as *cascading*, and several stages can be incorporated if desired. It means that we can reach quite low temperatures without having to build enormously strong and powerful compressors, or having to use a great deal of power at any one stage.

Having got a liquid gas at some low temperature (155° K (−118° C) for liquid oxygen, 20° K (−253° C) for hydrogen), we can obviously use it to conduct low-temperature experiments away from the machine, provided that we can keep it liquid. This is not quite as easy as it sounds. A vessel for containing liquid gases, and in which low-temperature experiments can be carried out, is called a *cryostat*. The inner experimental vessel is surrounded by another which contains the liquid, and the problem is to insulate this latter from the surrounding air. Liquid gases boil furiously at room temperature, and would boil away completely in a matter of seconds.

The essence of a cryostat, and the device which, perhaps more than any other, has made low-temperature research a practical proposition, is the Dewar flask, named after its inventor, Sir James Dewar (1842–1923). Actually, this piece of apparatus is familiar to everyone in a slightly different guise. The Dewar flask is quite as efficient at keeping heat in as keeping it out, and in the form of the vacuum or 'thermos' flask is an indispensable accompaniment at picnics. The flask in this case is inside a metal container which serves to protect it from shock and damage. The flask itself consists of a vessel with double walls; in effect, one flask within another. Between the two is a vacuum, so that heat cannot be gained or lost by the inner flask as there is nothing to conduct it. Loss or gain by radiation is reduced by silvering the inside surfaces. Thus with a Dewar flask it is possible to construct an efficient cryostat, and research at low temperatures can be carried out without too much difficulty, though it must be admitted that it is by no means easy, and requires considerable experimental skill.

Before we go on to look at some of the results of these researches, however, let us glance at some of the more practical reasons for liquefying gases. Not only liquefaction, but solidification of gases is possible if the temperature is low enough, and solid carbon dioxide, colloquially referred to as 'dry ice' is a very useful substance when it is desired to refrigerate something without the complication of machinery—when transporting ice cream or frozen goods, for example. But the principal use for liquefaction, and the reason why many tons of gases are liquefied every day all over the world, is in the separation of gases.

Both oxygen and hydrogen are in enormous demand for industrial purposes. They could be obtained by electrolysing water (that is, breaking it up electrically), but this would require vast amounts of power. There is plenty of oxygen all around us; one-fifth of the air is oxygen. It is also possible to produce hydrogen cheaply by passing steam over red-hot coke, the result being a mixture of hydrogen and carbon monoxide. This mixture, called water-gas, forms a fair proportion of the 'town's gas' which we use for everyday heating. The problem in each case is the separation. But if we liquefy air, it is found that the various constituents boil at different temperatures, and it is only necessary to collect them separately as the temperature of the liquid is allowed to rise slowly. The air is dried before being liquefied. At one time this was done in Linde Liquefiers, which used the nozzle expansion principle, but nowadays, oxygen is required in such vast quantities (it is common to speak of 'tonnage oxygen') for welding and steelmaking, that more efficient methods have had to be found and turbine liquefiers are now used.

The various gases 'distil off' as the temperature is allowed to rise. First of all, at 28° K, comes a gas called neon, which is responsible for the bright red glow in discharge lamps and tubes, such as are used for advertising purposes. Then at 78° K nitrogen comes off. There is not much use for this in the gaseous form. At 88° K, argon, a gas much used in certain welding processes, and also in 'gas-filled' electric light bulbs, distils. Then at 122° and 133° two 'rare gases', krypton and xenon, are obtained. These are only present in small quantities, but they also are used in certain kinds of electric lamps. Finally only the oxygen and the small quantity of carbon dioxide are left, unless the latter has been removed chemically before liquefaction to simplify matters. The oxygen can be boiled off and compressed into cylinders, but much of it is nowadays delivered to bulk users in liquid form.

The separation of hydrogen from 'water gas' is very similar. For another practical use of very low temperatures we can turn to a totally different field. In engineering, it is often necessary for one part to fit

159

firmly upon another, to which for practical reasons it cannot be welded, screwed, or otherwise fixed. In such cases extensive use is made of *shrinkage fits*. The fitting of a steel tire upon a locomotive wheel is a good example; the tire is made slightly *smaller* than the wheel centre and to get it into place it is heated, whereupon it expands sufficiently for the wheel centre to be dropped in. When it cools down, the tire is firmly fixed in place. Now it is not always possible or convenient to heat parts to the required degree. As we shall see in Chapter 7, many metals would have their properties altered by such treatment. However, cryogenic methods offer an alternative. Instead of heating the outer part, the inner part can be cooled, by immersing it in liquid air, for example. It will then contract sufficiently to be put in place, and when the assembly has warmed up to room temperature, it will be immovably fixed. This method has been used industrially for fixing liners in the cylinders of petrol engines. Incidentally, parts cooled in this way have to be handled with the same precautions as though they had been heated. A piece of metal at a temperature of 250° *below* the body temperature will cause a burn as serious as it would if it were the same amount *above* that temperature (see illustration on page 150).

We can now turn back to pure research. The lowest temperature so far mentioned is that of liquid hydrogen, at 20° K. Hydrogen was first liquefied in 1898, and with this success came the first indications that an entirely new realm of physics was being approached. Already at liquid-air temperatures, certain materials change their properties spectacularly: grapes frozen in liquid air will bounce like golf balls if dropped on the floor, whereas rubber completely loses its elasticity and will shatter like glass if struck. It can therefore not be used as a sealing material in cryogenic apparatus, and this is one of the engineering difficulties encountered at low temperatures. Another is that all ordinary lubricants freeze solid and lose their lubricating properties.

However, the changes noted at the temperature of liquid hydrogen are more fundamental. Measuring low temperatures is not easy; ordinary thermometers are useless, of course. Down to the temperature of liquid oxygen, thermocouples (see page 143) are satisfactory. However, when used in liquid hydrogen they gave indications which elementary theory showed to be far from accurate. Another method was tried, based on the fact that the resistance of a wire falls steadily as its temperature drops, but here again, the readings did not make sense. Eventually these low temperatures were successfully measured with an improved version of the old gas thermometer (see page 136) but it was obvious that strange things were happening as absolute zero was approached.

This realisation was the principal driving force behind the attempt to liquefy helium. This gas, mentioned on page 131, is a rare gas found in the products of certain natural gas wells in America; traces of it exist in the atmosphere. Next to hydrogen, it is the lightest of all gases, and has been used for filling balloons and airships as it is non-inflammable; in fact it is chemically inert. But the principal use for it today is in cryogenics. It was first liquefied in 1908 by Kamerlingh Onnes in Holland, and its boiling point turned out to be about $3°$ K. Since that time liquid helium has proved to be such a useful research tool that helium liquefiers were built commercially—they work on the expansion engine principle—and helium is now liquefied as a routine laboratory operation all over the world.

By allowing liquid helium to boil under a reduced pressure, temperatures down to $1°$ K can be reached, and this is the lowest that can be attained using liquefied gases. To get to still lower temperatures, other methods are needed. These are, first, magnetic cooling, by the aid of which temperatures down to about $0·003°$ K can be reached, and nuclear cooling, with which the lowest temperature so far recorded, $0·000016°$ K, that is less than two hundred thousandths of a degree above absolute zero, was attained at Oxford in 1956. Unfortunately, we cannot here go into details of these techniques, but it is to be hoped that the interested reader will study them in the appropriate literature.

It was mentioned above that curious things were happening as absolute zero was approached. At liquid helium temperatures, these phenomena are fully developed. One of the oddest is *superconductivity*. It had long been known that the electrical resistance of a metal falls with its temperature, and it seemed quite logical that at low temperatures the resistance should drop to low values. However, at liquid helium temperatures (the exact temperature varies with the metal concerned) the resistance of certain metals vanishes altogether. This at first sight seemed impossible, and even today it is not fully understood, but it undoubtedly happens, and there is some reason to think that at a low enough temperature it would happen in all metals. Superconductivity is usually demonstrated by inducing an electric current in a ring of a suitable metal. As long as this is kept below a certain temperature, the *transition point*, the current will continue to flow round the ring without diminution. The longest such a current has been kept going is two years, and there seems no reason why it should not be kept going for ever, provided that the supply of liquid helium could be kept up.

Almost as soon as the discovery of superconductivity was announced by Kammerlingh Onnes in 1911, suggestions were made for its practical application. Much of the weight, and the expense, of high-power electrical machinery is due to the very heavy copper conductors which are needed to carry large currents. Moreover, these large currents often give rise to a great deal of unwanted heat, and water cooling, with all the attendant complications, must be introduced. Technicians therefore began to dream of magnets and other devices with superconductive coils, taking up only a fraction of the space and weight occupied with conventional machines, and with none of the complications of special cooling systems. Unfortunately it was soon shown that the phenomenon of superconductivity vanished in quite modest magnetic fields, and so the dream faded.

It is only quite recently that superconductive magnets have become a reality; the wires used in them are made of special alloys which do not lose their superconductive properties at high magnetic fields. These alloys are made from rare and expensive metals, and the coils have to be surrounded by liquid helium. Superconductive magnets are used in cryogenic research itself, where very large magnetic fields are needed, and liquid helium will be used anyway. In these circumstances, the saving of copper, and even more important, the saving of electrical energy, make the cost of the special alloys worth while. Thus the huge water-cooled coils and powerful generators formerly used have given place to small compact coils and a few car batteries— a much more convenient set-up.

Recently (1968) the early dreams of superconductive machines have been brought appreciably nearer to reality. A British company has developed successful superconducting motors, and it is planned to install a large one, of 3,000hp, in a power station to test the principle thoroughly. It is expected to weigh less than one-eighth as much as the equivalent conventional motor, to cost little more than half as much, and to operate at about 97 per cent efficiency.

At the time of writing, much active research is going on with another practical application of superconductivity, and it is likely that by the time this book is published the first commercial installations will be working or very near that stage. This application is in computers. We saw in Chapter 2 that the design of powerful and compact computers is bound up with the problem of store, or memory, capacity. Superconductive elements are ideal for this purpose. They can store charges indefinitely and respond at any speed which is likely to be demanded of them, can be made very small, and require very little power. True, they must be kept at temperatures of around 4° K, but with the coming of commercial helium liquefiers, this is no longer a real problem. It seems entirely likely that the new generation of computers will make extensive use of superconductive stores.

Cryogenic research is today being conducted in many laboratories all over the world, and new discoveries are frequently reported. It is also certain that practical applications will be found for more and more of these discoveries. The space programme, with its need for huge quantities of liquid oxygen and, more recently, liquid hydrogen for rocket propulsion, has stimulated low-temperature engineering on an unprecedented scale. But for the scientist, the greatest attraction of this field is that it offers the opportunity of exploring an altogether different and unknown world. The properties of liquid helium itself are quite unlike those of any other substance; unfortunately lack of space precludes our going into them further. Before we leave the subject of low temperatures altogether, however, we should just glance at one discovery which may influence our everyday lives within a few years.

It was mentioned on page 143 that if the junction of two dissimilar metals is heated or cooled, an electric current will flow. This phenomenon, which is the basis of the thermocouple, is termed the *Seebeck effect*. It has long been known—since 1834 in fact—that there is an exactly analogous but opposite *Peltier effect* which causes a metallic junction to be heated or cooled if a current is passed through it, the direction of heat flow being determined by the direction of the current. This phenomenon remained a scientific curiosity since the heating or cooling effect was tiny, and in a metallic junction is masked by other effects. However, the development of the materials called semiconductors caused a revival of interest. The first, and best known use of these was in transistors, which have now virtually replaced thermionic valves (electronic tubes) in electronic equipment. The essential part of a transistor is a junction, but this junction does not involve separate pieces; it exists in a single piece of material. This is doubtless rather puzzling at first sight, but it is perfectly possible; the situation may be crudely visualised by considering the junction between the milk and cream in a bottle of milk. Furthermore, although the Peltier effect in metals is small, semiconductors can be made in which it is hundreds of times greater.

If, then, a semiconductor junction be made in one of these materials, and a current passed through it, one side of the junction will be heated while the other is cooled; in other words, heat is made to flow across the junction. The effect can be enhanced by compounding the material so that its own thermal conductivity is low, and there is very little reverse leakage of heat. A practical device utilising this principle is made up of many such junctions; if one end of this is built into a suitable enclosure, the other end being open to the air, the temperature of the enclosure can be reduced by about 25° C, that is, from normal room temperature to below freezing point. By adding a second stage, a greater degree of cooling is possible. These devices have already been applied to such purposes as instrument cooling, where their small size and lack of moving parts are great advantages. At present, the capital cost is too high to allow them to compete with conventional refrigeration methods in less specialised fields, but if the cost could be sufficiently reduced, as it well might be by improved manufacturing techniques and quantity production, it is obvious that here is the ideal basis for a domestic refrigerator. There are no moving parts, nothing to get out of order, and the size and weight are small. All that is needed is an electrical connection and a thermostat. It seems safe to predict that much more will be heard of thermoelectric cooling in the future.

CALORIMETRY

It will be as well to bring this chapter to a close with a brief explanation of the units used to measure heat, as opposed to temperature. The two are, of course, linked, and the methods of measurement depend upon this. Using the British system of weights and measures, the unit of heat is the British thermal unit (Btu), which was defined on page 94 as that amount of heat needed to raise one pound of water through one degree Fahrenheit. In the cgs system, the unit is the calorie, defined as the quantity of heat necessary to raise one gram of water from 15° C to 16° C. The exact temperatures are specified because the amount of heat required to raise the temperature of water by a given amount (its *specific heat*) varies slightly with the exact temperature. In the MKS system, the unit is the Calorie, which is the quantity of heat required to raise one kilogram of water through a similar temperature range. Note that the only difference in the manner of writing these two units is in the use of the capital C for the second one. In many ways this is inconvenient, and can be confusing if the printer is careless, but it is established. Probably, as the MKS system displaces the cgs, which it seems likely to do, the difficulty will resolve itself; meanwhile, one must be alert to the difference when reading scientific literature.

The definitions of these units give a clue to the methods of measurement. Suppose it is desired to measure the calorific value of some fuel. The fuel is enclosed in a sealed container with a quantity of oxygen sufficient to ensure its being completely burnt. This container is completely immersed in a known quantity of water, usually for purposes of easy calculation, one litre or a multiple thereof, or an imperial gallon if the British system is being used. The outer vessel containing the water is carefully insulated, so that heat is not lost. The entire set-up is called a *calorimeter*. The fuel is now ignited, which is done electrically, and when it has completely burnt, the temperature of the water is taken. Thus the quantity of heat produced, in terms of calories, Calories, or Btu, can be calculated, and this, related to the weight of fuel, gives the calorific value. The heat produced in a chemical reaction, or an electrical process, can be measured similarly.

6

Very Small

It is a matter of common experience that unaided eyesight is not good enough to find out many things. Many important discoveries and developments depend on the ability to supplement the human eye with lenses and microscopes. The manufacture of glass lenses was probably first suggested by observation of the properties of a glass globe filled with water. Who exactly was responsible is not certain, but lenses are mentioned in the writings of some of the tenth-century Arab philosophers, and they were known at the beginning of the thirteenth century to Robert Grosseteste and his better-known pupil Roger Bacon, who knew of their magnifying properties and suggested their use in spectacles. The first actual spectacles were made in Italy towards the end of the thirteenth century. They used convex lenses to compensate for long sight and to magnify.

The illustration, Fig 20, shows how a convex lens magnifies. If a ray of light passes at an angle from a medium such as air into a denser one, for example water or glass, it is *refracted* (bent) and the same thing happens when it passes from the dense to the less dense medium, only this time it is bent in the opposite direction. If the piece of glass is of a suitable shape, all the rays of light which fall upon it will be so refracted that they meet at a certain point, which is the *focus* of the lens. This fact is well known to anyone who has used a lens as a burning glass. If one looks at some small object through a lens, the path of the rays of light will be as shown in the figure. The rays from the object will be refracted as shown, but the eye (or more accurately, the brain) perceives the object as though these refracted rays had come straight from it, and it is therefore seen as much larger and somewhat farther away than it really is.

A little consideration of the diagram will show that the greater

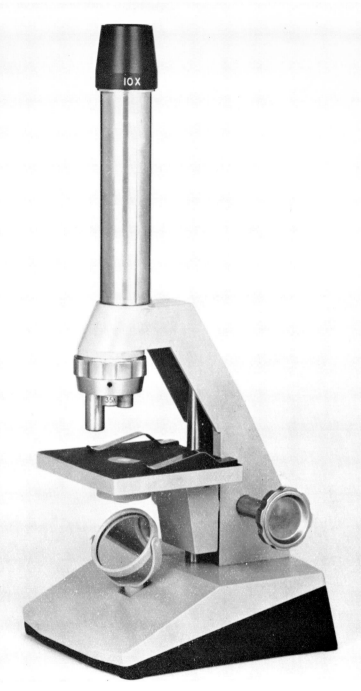

Page 167 The Beck 'Student' microscope. Specially designed for use in schools, it is a noteworthy example of modern industrial design. As shown here, it is in the upright position; it can be used in an inclined position by simply reversing the lower part of the base on a swivel. An integral lighting device can be used in place of the mirror shown

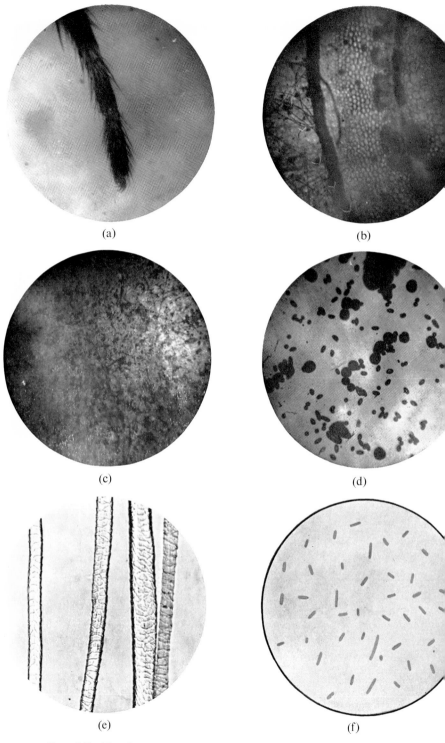

Page 168 For descriptions refer to text, pages 176–9. Magnifications: (a), × 20; (b), × 30; (c), × 50; (d), × 50; (e), × 200; (f), × 2,000

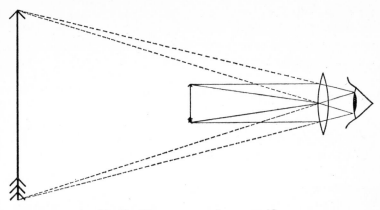

Fig 20   How a convex lens magnifies

the curvature of the lens, the greater the total refraction and thus the magnification. The first microscopes had lenses that were almost spherical, being made from fused glass beads. They were not particularly efficient, for they gave distorted images, and it was not until the Dutchman Antonij van Leeuwenhoek (1632–1723) perfected a method of grinding the lenses to shape that much progress was made in microscopy. With his lenses van Leeuwenhoek was able to see the minute life forms in stagnant water, which are quite invisible to the naked eye.

Even with improved techniques, it is not a particularly easy matter to produce a near-spherical lens; fortunately the same result can be obtained by combining two or more convex lenses so that each one contributes its quota to the total refraction, and this is the principle of the compound microscope. This was already known before the time of van Leeuwenhoek, and the first compound microscope seems to have been made in about 1590 by another Dutchman, named Zansig. Such an instrument was also described by the English scientist, Robert Hooke, about 1650.

In principle, a compound microscope can magnify much more than any single lens, but the early ones suffered from the disadvantage that the faults of the primitive lenses—called *aberrations*—were magnified like everything else. Not until the causes of the aberrations were understood was it possible to produce lenses good enough to make

L

the compound microscope a really practical proposition. From then on progress was steady, if unspectacular, and some remarkable results were obtained with apparatus which by modern standards was incredibly primitive.

THE MODERN MICROSCOPE

The microscope as we know it today, however, did not come into being until the famous collaboration of Abbe and Zeiss. Carl Zeiss was the mechanic and instrument maker to the University of Jena. One of the most brilliant mechanics of his day, he had already begun to improve the design of the microscope, and to build instruments according to his own ideas, when in 1866 he first met Ernst Abbe, already at the age of twenty-six a professor of physics in the university. This meeting was to lead to an epoch-making partnership, which is noteworthy as being one of the first deliberate applications of science to industry. Zeiss's great difficulty in constructing his improved microscopes was in grinding the lenses. He was obliged to proceed according to the method, usual at the time, of trial and error; scarcely two instruments were identical, and getting a good one was mainly a matter of luck.

Abbe and Zeiss seem to have got on well from the first, perhaps because Abbe himself was from a working-class background, an unusual thing in the academic circles of nineteenth-century Germany. Anyway, Zeiss was soon discussing his problems with Abbe, and before long he suggested that Abbe should undertake the task of putting the design of microscopes on a scientific basis. It was a problem after Abbe's own heart. For four years he studied it and worked at it, until, in 1870, he presented Zeiss with the results of months of calculation, which not only put the manufacture of lenses on a firm scientific and mathematical basis, but which amounted to the designs for an entirely new instrument. Practically speaking, the general-purpose microscope of today is the instrument of Abbe and Zeiss, with but few and minor improvements. The firm of Carl Zeiss became world famous, and it is pleasant to recall that Abbe had his just reward; Zeiss made him a partner in the undertaking. Let us now see how this instrument works, and what it can do.

A modern general-purpose microscope, as used by students, is shown on page 167, and its working can be understood with the aid of the simplified diagram in Fig 21. Light, either daylight, or better, light from a specially designed lamp, is reflected from the mirror through a system of lenses, the *condenser*, which serves to concentrate the light on the specimen to be examined. This is mounted on a glass slide, and must, with the arrangement being described, be translucent. The slide rests upon the *stage*, being held by spring clips. The light then passes into the optical system of the microscope proper, which consists of two distinct parts, the *objective* and the *eyepiece*, each of which contains at least two, and often more, individual lenses. They are mounted at the foot and top of the microscope tube, in such a way that they can easily be changed if need be. Either the tube or, as in the instrument shown, the stage, is raised or lowered to bring the specimen into focus; coarse and fine adjustments are provided for the purpose in the more elaborate instruments. The diagram shows the path of the light rays within the instrument, and clearly illustrates how the enlarged image formed by the objective is, so to speak, 'picked up' and still further magnified by the eyepiece.

The total magnification is found by multiplying together the magnifications of the objective and eyepiece, and since both objectives and eyepieces are obtainable in various 'powers', a combination to suit the work in hand can easily be chosen. The microscope shown has a *triple nosepiece*, which allows any one of three objectives to be brought into use in a few seconds. Eyepieces are equally easy to change. An instrument such as this would be provided with objectives and eyepiece giving magnifications of from about 35 diameters (written as × 35) to some 200 diameters. The more advanced microscopes used for such purposes as bacteriological research are capable of magnifications of up to × 2,000, but here it should be said that magnification alone is not the only thing—indeed it is not the principal thing—that matters. What is known as *resolution* is just as important, and this depends upon the quality of the objective. The resolving power of a lens is its ability to separate fine detail, and no amount of magnification will be of any use once the limit of resolution has been reached. It is, in fact, referred to as 'empty' magnification. High-quality modern objectives can resolve detail as small as $0 \cdot 2\mu$ ($0 \cdot 000008$in), and this is very near the theoretical limit, to which we shall return later.

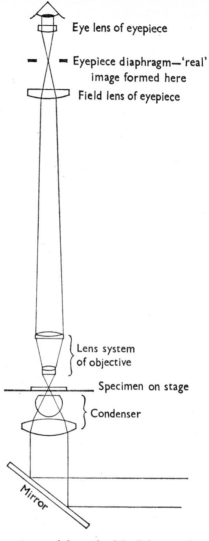

Eye lens of eyepiece

Eyepiece diaphragm—'real'
image formed here

Field lens of eyepiece

Lens system
of objective

Specimen on stage

Condenser

Mirror

Fig 21 The lens system and the path of the light rays through a compound
microscope

All optical microscopes work upon the principles just described, and many thousands of instruments of a kind generally similar to the one shown are in use in laboratories all over the world. For some purposes, however, it is convenient to have a microscope specially designed for the job, and there are several types of such specialised instruments. The high-power instruments used for bacteriology have already been mentioned. The distinguishing features of these, apart from the very high quality of the optical equipment, are the solidarity and accuracy of their construction. In many machines, 'play' of 0·001in is considered to be quite acceptable. Most cars which have been running for any length of time show a lot more than that in most of their parts! But in a microscope, magnified by 1,000, such a play, or shake, would become one inch. Try reading a book while shaking it about to that extent! For practical purposes, no play whatever can be allowed in a high-power microscope; moreover, none must develop during the life of the instrument. This naturally means that special techniques of construction must be employed, and such instruments are expensive. A standard feature of such a microscope is a 'mechanical stage', which enables the operator to move the specimen slide by very small amounts under close control, a great help when systematically searching for bacteria and the like.

Petrological microscopes are used for examining mineral specimens. They are always fitted as standard with the apparatus for making examinations by polarised light. The theory of this is briefly discussed in Chapter 8; it is a technique which is very useful when examining minerals and crystals of all kinds. Also, the stage of these microscopes is made so that it can be rotated. So far, we have assumed that specimens are translucent, but of course, this is not always the case. Opaque specimens can be examined quite well at low powers by directing a beam of light on to the surface, but a high-power objective has to work close to the specimen, and tends to get in the way. Moreover, for many purposes it is desirable that the light be projected vertically on to the specimen. High powers and vertical illumination are essential when examining metal samples, and metallurgical microscopes have special illuminating arrangements. The metallurgical microscope, and the technique of examining metal specimens, are described in greater detail in Chapter 7.

Any of these more advanced instruments can also be made as

binocular microscopes. With the more usual kind of microscope, only one eye is used, of course, the vision of the other one being mentally ignored. The unused eye is *not* closed; this would be courting a headache. The technique of working with both eyes open is not nearly as difficult as it sounds; nevertheless, using only one eye for long periods is attended by a certain amount of strain, even if both eyes are used alternately, as all expert microscopists do. The binocular microscope overcomes this; it has twin eyepieces and a prism system, so that both eyes can be used at once. Such an instrument is not cheap, but in a pathological laboratory, for example, where a worker may spend the greater part of his day at the microscope, it is well worth while. It has the additional advantage of giving stereoscopic vision.

This latter feature is also valuable in the dissecting microscope, which is a low-power binocular instrument used, as its name implies, for dissecting small objects and organisms ready for examination by a larger instrument. This kind of microscope also has an extra lens system for erecting the image; the image produced by an ordinary microscope is reversed and inverted; this does not matter for most purposes, but it would be a nuisance in dissecting.

Finally, portable microscopes are made, and a great deal of ingenuity has gone into making them as small and light as possible, while retaining many of the features of a full-size instrument. The latest development in this field is a microscope small and light enough to be held and used in one hand, but which is capable of a performance equal to that of the best orthodox instruments. In fact, it can do a number of things which the latter cannot, and may well supplant them in the future. This result has been achieved by an ingenious 'folding up' of the light paths through the microscope.

PHOTOMICROGRAPHY

Microscopists have always desired to have a permanent record of the things which they have seen. To some extent this desire can be met by 'mounting' specimens permanently, that is, by so treating them that they are preserved and fixed upon slides for future reference. Many objects—bacteria for instance—often require a certain amount of such treatment before they can usefully be observed at all, but the technique, important though it is in microscopy, has limitations. Living organisms are necessarily killed by it, and may be distorted in the process. Very great skill is needed for some kinds of mounting; by no means everyone can do it. And of course a mounted slide has distinct limitations for teaching purposes, and is useless for illustrating a book or paper.

In earlier days, reliance had to be placed on drawing. Some of the older textbooks contain beautiful examples of such drawing, often lithographed in colour. Not everyone, however, has the artistic talent necessary to do this sort of thing. The situation was eased a little by the invention of the *camera lucida*, which is a device which enables the observer looking into the microscope to see the image of the specimen as though it were projected on to a sheet of paper placed on the table in front of him. A record can then be made by tracing over this image. Even with this aid, a fair degree of skill is required, but there is another, more fundamental, objection to drawing as a means of record, and that is the tendency of the artist to select. This can be most valuable, as when indicating to a student the essentials which he should look for amid a mass of irrelevancies, but it can also be dangerous, for the human brain has a tendency to simplify and impose order upon the outside world. Many of the things seen through the microscope are complicated in the extreme, and this tendency may lead to important matters being omitted from the drawing, or worse still, distorted in accordance with the observer's own (often unconscious) preconceptions.

The invention of photography provided the answer to this problem, and nowadays all really important or interesting observations are photographically recorded as a matter of routine. It is perfectly possible to take a photograph through the microscope with an ordinary camera (I have done it more than once), but for serious work it is customary to use either a camera so made that it can be conveniently linked to the microscope, or a piece of apparatus specially designed for the job, and consisting of camera and microscope in one.

A photograph taken through the microscope is called a *photomicrograph* (*not* a microphotograph, which is something quite different), and examples of a few are shown on page 168. These give some idea of what can be seen and discovered with the microscope, but the applications of microscopy would require a book just to list them, and only these few typical ones can be briefly described here. Several examples of the practical utility of the microscope will, however, be found in the following chapters.

Photograph (a) shows the leg of a common garden spider. Notice the delicately formed claws at the foot. These enable the spider to run about over its web, and also have a function in the spinning and preparation of the web. This is a good example of the kind of thing which can be seen with quite low powers and simple apparatus, but which could otherwise only be guessed at. Without the microscope, the detailed study of the anatomy of spiders, insects, and similar creatures, with all its important implications in pest control and agriculture, would be virtually impossible.

The next photograph, (b), is a botanical specimen, a cross-section through the stem of a rose. The microscope has shown us that all forms of life, with the exception of certain very simple ones such as viruses, of which more later, is organised in the form of cells—from a single one in the lower forms to many millions in the higher plants and animals. This photomicrograph illustrates this very clearly. The large cells, on the right of the picture as printed, are actually in the centre of the stem. To the left of these can be seen the groups of *vascular bundles*, which are cells elongated into conducting tubes. A plant is, in a sense, a complex chemical factory, producing all kinds of substances from the basic raw materials of carbon dioxide (from the air), water, and various mineral salts (from the soil). These conducting tubes carry the raw materials and finished products to and from the leaves and roots. A further interesting feature of this micrograph is the fibres seen projecting from the outer surface of the stem on the left. These are threads of the rose mildew. The microscope does, of course, play a most important part in investigating diseased conditions in both plants and animals. In order to see cell structure properly, sections of no more than a few thousandths of an inch in thickness, and often much less, are needed. These are cut with the aid of a device called a 'microtome', which may be no more than a hand-held gadget to guide the blade of a razor, but which for advanced work can be an elaborate machine capable of cutting 'serial sections' automatically and with great accuracy.

We turn next (c) to something very different and seemingly less interesting—dust. But a glance at the picture shows the unsuspected detail and diversity which the microscope can reveal in even the most humdrum objects. Dust consists of tiny particles, most of which are detached from larger masses by the natural process of wear. Thus there may be grains of soil, rock particles, wood particles, small pieces of metal, fibres, and much else. The interesting and important thing here is that the dust taken from a given place is almost certain to be quite specific to that place, since (to take an everyday example) the construction, furnishing, and uses of a room are unlikely to be exactly duplicated anywhere else. The main practical application of these facts is in forensic science, where the microscopic examination of dust is an important study. An expert in this field can identify an astonishing variety of dust particles simply by their appearance, and tests have been developed for distinguishing between particles of similar appearance. A visitor to any locality is quite likely to carry away with him some of the dust associated with that locality or with something which has happened there; thus, safe-breakers have been convicted because dust particles found in their clothing were shown to have come from the fireproof lining of a certain safe—thrown unnoticed into the air when it was blown open by explosives. In fact, that particular branch of 'finding out' exemplified by detective work owes a great deal to scientific methods in general and to microscopy in particular.

One of the things which can be found in all dust is pollen grains. Our next example (d) shows a selection of these, taken from different plants, and it will be seen that they are highly individual. This fact in itself is useful in the examination of dust, for grass pollen will not be found in quantity in woodland surroundings, for instance, and the air and therefore the dust of a town will not contain pollen associated with cornfields. But the importance of pollen analysis goes beyond this, which is why it is introduced here. It has, in fact, become a study of sufficient interest to be dignified by a name of its own—palynology.

The thing that makes this possible is the fact that the material of the outer coat of a pollen grain (the exact composition of which is as yet not fully understood) is one of the most stable and resistant of all of the substances produced by living organisms, and under favourable circumstances it may be preserved in a recognisable form for thousands of years. Palynology is strictly the study of fossil pollens of

177

this kind. Thus, if some remains which can be definitely dated are uncovered in the course of archaeological excavations, and if the soil around them is shown to contain fossil pollen grains from, shall we say, the pine tree, then we can assert with some certainty that these remains were associated with an environment of coniferous wood-land, even though today the land is given over to agriculture or meadowland. Further, knowing this, we can use the fact to date similar remains found in the same surroundings.

Pollen analysis is playing an increasingly important part in archaeology, and it has shown beyond doubt that the vegetable cover, and therefore presumably the climate, has changed drastically in some parts of the world even within the past few thousand years. A rather more extreme example is interesting as it shows how long these fossil pollen grains may persist under some circumstances: in Vermont, one of the north-eastern states of the USA, there are deposits of lignite (brown coal) dating back to the Oligocene period, about 17,000,000 years ago. These have been carefully studied, and the remains of more than 56 genera of flowering plants have been identified. Some 60 per cent of these are represented by pollen alone. The mixture of plants is quite unlike that associated with present-day Vermont, but is closely similar to that found in the river swamps of Florida and South Carolina, showing that at one time the climate of Vermont must have been rather like the sub-tropical climate of these places today.

Fibres were briefly mentioned above, when we were discussing dust, but the microscopic examination of plant, animal, and man-made fibres is of great interest and importance in itself, particularly in the paper and textile industries. In fact, whole books have been written about it. The example shown here (photograph (e)) is of wool fibres. Notice the scaly structure of the fibre surface. This is one of the things which gives wool its unique properties, and the tremendous difficulties of imitating this structure artificially is one reason why wool can maintain its position as a textile fibre in the face of competition from the man-made fibres.

So far we have been looking at more or less everyday objects, and we have seen how the microscope can reveal previously unknown and unsuspected aspects of them. For our last example, however, we turn to a world the very existence of which would be unknown but for the microscope. The photomicrograph (f) shows bacteria—minute, single-celled life forms which exist all around us. The ones shown here are *Escherichia coli*, which inhabit the human digestive tract. The average man has several millions of these inside him, 'peacefully co-existing', so to speak. Bacteria are the causal agents of many diseases, and their study is an important branch of medicine, but these 'pathogenic' bacteria are only a tiny minority of the known species, of which there are now many thousands. Bacteriology is a science in itself. The microscope is its principal tool, but all kinds of ingenious techniques and equipment have been developed for collecting, cultivating, preserving, and separating bacteria, and for preparing them for examination (usually done by staining methods). Even a brief popular treatment of the subject calls for a book to itself, and the interested reader is recommended to some of the books in the Reading List.

THE ELECTRON MICROSCOPE

It was mentioned above that there is in fact a theoretical limit to the degree of resolution possible with an optical microscope. This limit arises from the nature of light. As we saw in Chapter 3, light is a wave phenomenon, and the various colours correspond to various wavelengths. The limit of resolution of the optical microscope was calculated by Abbe; we shall not go into his arguments here, but they lead to the following result:

$$h = \frac{0.61\lambda}{\text{NA}}$$

in which formula h represents the smallest distance between two objects which can be resolved, λ stands for the wavelength of the light being used, and 'NA' is an abbreviation for 'numerical aperture', a figure which depends upon the design and construction of the objective lens; for the best high-power modern objectives it is around 1.40. In other words, the smallest object or detail which can be seen is approximately half the wavelength of the light being used. If the reader has fairly strong powers of visualisation, he may be able to see this for himself intuitively, without any calculations, but in any case the fact is amply confirmed by experiment. Sometimes, in order to get the best out of a given microscope, it helps to use blue light, which has a short wavelength.

179

Very Small

One way of extending the limit of resolution of the orthodox type of microscope is to use ultraviolet radiation, which has a shorter wavelength than visible light. The human eye is insensitive to this radiation, of course—in fact, it is damaged by it in any large amount, hence the necessity for dark glasses when taking 'sun-ray' treatment, which is very rich in ultraviolet. Fortunately, however, this kind of radiation affects a photographic plate very strongly, so we can take photomicrographs. Glass is somewhat opaque to ultraviolet, and lenses for use with it are made from fused quartz, but apart from this the principles of a microscope for use with ultraviolet are much the same as those of one for visible light.

This pushed the limits a little further, but it was at best a refinement; much remained beyond the power of the microscope, and in particular it was impossible to see the viruses, minute life-forms which were known, from other evidence, to exist, and a knowledge of whose structure was vital to the advance of medicine. The answer was found at about the time of the Second World War. It is the electron microscope, and we must now, to bring this chapter to an end, briefly see what it is and what it can do.

Electrons are one type of the 'elementary' particles which make up all matter; they are thus much smaller than atoms. In fact they can be regarded both as incredibly tiny particles or as waves. This dual nature (which the reader is asked to accept for the moment) is very useful. Electrons can be produced and directed electrically (focused in fact) just as though they were a beam of particles (this is done in the cathode-ray tube of a television receiver), but this beam can also behave, in some circumstances, as though it were a beam of radiation of (and this is the important point for us) extremely short wavelength; much shorter than that of light.

The method of applying this practically in the electron microscope is shown in Fig 22, which should be compared with the photograph of the instrument on page 185. The electrons are produced by the electron gun—which, put very simply, is just a piece of hot wire—at the top of the upright tube. They are focused on to the object by the magnetic lens system, which corresponds to the condenser in an optical microscope, and the image is formed and magnified by further magnetic systems which are the exact equivalents of the objective and eyepiece. There are certain practical limitations, which for a

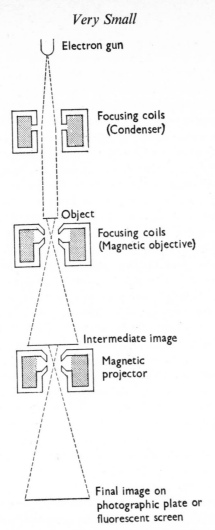

Fig 22 The principal components, and the path of the electron beam, in an electron microscope

long time made electron microscopy a very difficult and specialised business. To start with, the image is quite invisible to the human eye, and it has to be made visible either by using a fluorescent screen (again, not unlike the one in a television set) or by photography.

Strictly speaking, it is incorrect to refer to *photo*graphy in this connection, and the pictures taken with an electron microscope are called *electron micrographs*. Another snag is that electron radiation is not very penetrating; it is stopped within a few centimetres by ordinary air, and denser substances offer an even greater barrier. This means that the whole of the working portion of the microscope must be completely evacuated; moreover, specimens cannot be mounted on glass slides, and special methods of mounting have had to be devised.

Unfortunately, the electron microscope is not complete in itself; it requires supplementary equipment in the form of high-tension electrical apparatus, control circuits, and extremely efficient vacuum pumps of advanced design. All this tends to make it a somewhat expensive instrument. Formerly it was no light thing to break the vacuum inside the tube, and elaborate systems of airlocks were provided for inserting and removing specimens, but modern instruments have been reduced in size, and with efficient pumps, the working vacuum can be reached in a very short time, which greatly simplifies the operation. The magnetic lenses have also been considerably improved. The early electron microscopes, while they represented a tremendous advance on the optical microscope, still fell a long way short of the theoretical possibilities, and this was due in great measure to the imperfections of the magnetic lenses. In fact, the electron microscope has been enormously developed since it was first introduced. It is now a standard laboratory instrument, though still a somewhat expensive one. Preparation of specimens, at one time a very complicated and tricky business, has been greatly simplified, and we are now within sight of being able to examine living objects as an everyday routine. The latest development, the scanning electron microscope, makes possible the examination of much larger specimens than hitherto, and also gives a degree of resolution and definition which represents a tremendous advance on anything that has gone before.

With all this, however, the electron microscope has not supplanted the optical microscope, nor is it likely to do so. Rather it takes over where the optical microscope leaves off. Even at this, it has opened up a whole new world to us, and an excellent example of the sort of thing it reveals is shown on page 186.

7

How Strong is it?

The story has often been told of how, in a certain Cambridge college some time before the Second World War, a toast was drunk with considerable enthusiasm: 'Here's to mathematics and pure science, and may they never be of any damn' use to anybody!' This, of course, can never be more than a pious hope. It is unquestionable that many (though by no means all) of the most fundamental discoveries have been made as the result of a disinterested pursuit of knowledge for its own sake; the idea behind the toast seems to be that such discoveries *ought not* to be put to practical use, an idea which still lingers in places. This, however, is not the view of most of us today. So now, having looked at the tools of research, so to speak, let us see what can be done with them in the more mundane world of the workshop.

We shall first look at the testing of materials. This is a comparatively recent development. The materials of construction used in the past were of very variable quality, which was only to be expected, since many of them, such as stone or wood, were used in their natural state. Even manufactured materials depended very largely upon the quality of the raw materials for their final properties. Such things as metals, glass, pottery, and textiles fall into the latter category, and without a means of controlling the quality of the raw materials, any but the crudest forms of testing of the finished material would have been a waste of time. However, most such materials were cheap and plentiful, and the custom was to make structures 'amply strong', as it was usually expressed. If the materials happened to be good, the structure might last for hundreds, even thousands, of years, and we can see examples all around us in old buildings and various other relics of the past.

Of course, not everything lasted, even if built upon these principles, and there must have been a good many errors of judgement: thus, the supporting pillars of the dome of St Paul's cathedral were not adequate to their task, and the dome might have fallen by now had they not been strengthened by modern techniques. Still, the failure of a structure was not usually disastrous.

We cannot afford this approach today. Modern materials are often very expensive, and it is uneconomic to ensure strength by the crude method of using more material than is necessary. Furthermore, the consequences of failure in many modern structures would be catastrophic. The matter first became of importance with the Industrial Revolution, and the increased use of machines and (comparatively) high-speed means of transport. One of the earlier instances of the failure of a large engineering structure was, unhappily, attended by considerable loss of life: this was the Tay Bridge disaster in December 1879, when 73 persons were drowned. Nowadays, a single air accident might well have a similar death roll, the loss of a large liner or the collapse of a modern high building would be worse, while the consequences of the failure of a nuclear reactor would be unthinkably terrible. To a large extent, of course, avoidance of failures is a matter of design, but even the best designs are useless without proper materials, and indeed, in many cases design cannot even be started without accurate data on the materials. We must *know* the properties of our materials before we use them.

Rough-and-ready tests of materials have, of course, been made from time immemorial, and are by no means superseded even today, though many of them yield but little information, and that of doubtful value. Most such tests depend upon the skilled judgement of the experienced worker. Thus, in an old book on building construction, which belonged to the present writer's grandfather, there appears, beside descriptions of rather more scientific procedures, the recommendation to make an on-the-spot test of a consignment of Portland cement by thrusting one's bared arm into it up to the elbow! It is alleged (I have never tried it) that much could be learned from the general 'feel' of it against the skin.

In much the same way, a builder will even nowadays take a few bricks from a consignment and assess the general quality by visual examination, and perhaps by breaking a few of them to see what the

Page 185   The RCA model EMU–4 electron microscope

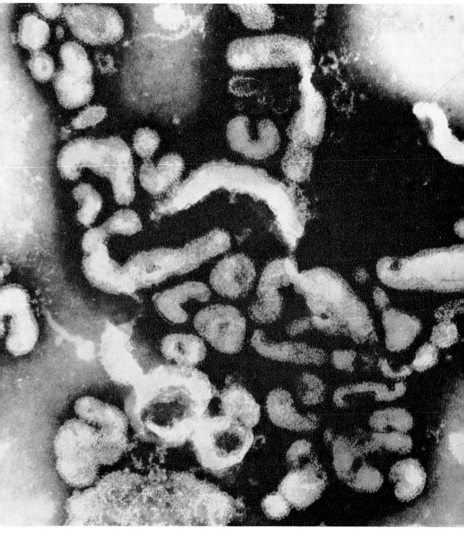

Page 186 An electron micrograph of influenza virus, taken by Dr Ward R. Richter of the University of Chicago, using an instrument of the type shown on the previous page. Magnification of the original × 120,000; here somewhat reduced for printing

inner structure is like. Metalworkers have for long been accustomed to do the same, breaking a small sample and attempting to gauge its quality by examining the appearance of the fracture. A subjective judgement of the force required to break the specimen would also be made. It is not without interest in this connection that the very word 'test' comes from the Latin, and meant originally the small crucible in which these samples were prepared.

A more important aspect of these tests is their assumption that 'the sample is representative of the bulk', and this principle is indeed central to most testing procedures. Contracts for the supply of materials usually include a clause to the effect that a sample of a stated size shall be truly representative of the bulk. Obviously, it is often impracticable, and may be impossible, to test the whole of a consignment—apart from the time factor, tests frequently involve the destruction of the sample. The question of sample size, and the deductions which can reasonably be drawn about the bulk from the sample behaviour, are part of the province of statistics, which was briefly discussed in Chapter 2.

STRENGTH OF MATERIALS

Before we consider tests of the strength of materials, it is necessary to look a little more closely at the term to see exactly what we mean by it. 'Strength of materials' is a recognised branch of engineering science, and the subject of many specialised books and articles, so it is fairly obvious that there is more in the matter than immediately meets the eye. In this section we shall consider the various aspects of strength, and introduce one or two technical concepts necessary to a proper understanding of the matter.

Nearly all the materials most interesting from a practical point of view are either crystalline in nature or they are *polymers*. Crystalline structure is not always immediately apparent to the naked eye, but it can usually be detected by the microscope, and this shows that, for instance, all metals are crystalline. A polymer is a substance the molecules of which are very large, and are built up from a great number of simpler units linked together. Most of the newer 'man-made' materials are polymers, and a whole branch of chemistry has grown up around them, but there are natural polymers as well, and

M

in particular all the traditional textile fibres belong in this group. Crystalline materials are generally composed of a large number of small crystals—a large single crystal is quite a difficult thing to produce, though it can be done, and is necessary for some purposes. Polymer molecules are put together in a definite way, often being longer than they are broad. The most important consequence of these facts is that materials are seldom equally strong in all directions or for all the possible ways of loading them.

Thus, a cast-iron* column can easily support a heavy roof, for example. But if the column were to be suspended by one end, and then loaded by hanging a weight equivalent to that of the roof on the other, it would almost certainly break. We say, therefore, that cast iron is strong in compression but weak in tension, or, using the terms generally encountered, that its *compressive strength* is high but its *tensile strength* is low. Again, if cast iron is placed in a situation where it is subjected to forces which tend to push different parts in opposite directions (a bolt preventing two plates from sliding over one another is an example), it will fail at a very low load. It is said to have a low *shear strength*, and steel would always be used in such a situation.

It may be that the material is required to form a moving part in some mechanism. In that case its tensile, compressive, or shear properties will be arranged to be suitable for the job in hand, but it may also be important that it should have the maximum resistance to wear. In that case its most important property may well be its *hardness*. This is a property not immediately connected with what we should normally define as strength, but it is nevertheless an important one. There are several other such, some of which have come into prominence only recently.

Thus, if a lead wire be taken and loaded with a weight well short of that required to break it, nothing very obvious happens at first. If the weight is immediately removed, no change in length is detectable upon measuring the wire. But if the weight is left in place for some time, it is found that the wire has stretched. This phe-

---

* For descriptions of the various kinds of iron and steel, details of their manufacture, and a certain amount of information concerning tests during manufacture, consult *The British Iron and Steel Industry, A Technical History*, by W. K. V. Gale (David & Charles, and Augustus M. Kelley, NY).

nomenon is known as *creep*. It can be seen in lead, tin, and zinc in the manner described, but fortunately the main structural metals, iron, copper, and their alloys do not show it at ordinary temperatures. They do exhibit creep at high temperatures, however, and this is of great importance in gas turbines, for example. In fact, a great deal of work has gone into the development of *creep-resistant alloys* for this and similar purposes.

We thus see that the strength of material is not necessarily a constant factor, but may vary with the changing conditions. One instance of this which was brought into public prominence a few years ago, although it had long been known to engineers, was the question of *fatigue* strength. You can demonstrate for yourself what this means as follows: take a piece of tinplate (which can be cut from a tin can with an old pair of scissors—but make sure there are no sharp edges), clamp it by one end, or nail it to a post or bench top if you haven't a vice, and try to break it by pulling on one end with a pair of pliers. You will most certainly fail. Now bend the strip back and forth, doubling it upon itself about a dozen times, and try once more to break it. This time it parts quite easily; indeed, it may break without your having to pull it at all, if the bending is kept up long enough. The metal is said to have become *fatigued*, and this is always liable to happen when metals are subjected to repeated loads over a long period of time, even though no individual load would be sufficient to cause failure.

Many materials can be damaged and weakened by radiation of the kind produced in nuclear reactors, and tests have had to be devised so that the best materials for working under these conditions can be selected. Even in less dramatic applications, the effect of the working environment on materials may have to be taken into account. Luckily, most metals are unaffected by water, but the same cannot be said for all materials, and nobody in his senses would build a boat using animal glue in the joints. *Wet strength* can be an important limitation on the use of organic materials.

The statement that water does not affect metals may be disputed on the ground that iron and steel are known to rust badly under moist conditions, but in fact the chemistry of this situation is more complicated than appears at first sight, and both air and water are necessary for rusting to take place. Iron does not rust in dry air; neither

189

does it if it is immersed in water which has been boiled to remove all the air.

Rust is only a specific example of the general phenomenon of *corrosion*, of course, and it may be necessary to apply tests to make sure that a metallic part will not fail due to this in the working environment for which it is intended.

These, then, are some of the properties of materials which we may want to know about, and we will now see how we can find out about them.

EXAMINATION OF STRUCTURE

It was pointed out above that many materials, and in particular metals, are crystalline. Much can be ascertained about a material by an informed examination of its crystal structure, which depends first on its constitution, and second, on the treatment which it has undergone. Thus, if a piece of tool steel is heated to redness and allowed to cool slowly, it will be quite soft, and can be filed, sawn, or machined. In this condition it is said to be *annealed*. If it is now again heated to redness, but this time cooled quickly, as by plunging it into cold water, it will be hardened, so that it can itself cut another piece of steel. The hardness can be regulated by heating it to varying temperatures, less than the original red heat, and once more cooling. These varying degrees of hardness and softness are reflected in the crystal structure of the metal as seen through a microscope.

To prepare a piece of steel, say, for microscopic examination in this way is quite a simple matter, though it demands a good deal of patience. The specimen (as the sample to be examined or tested is usually called) is first of all filed or ground flat and smooth. Care has to be taken when doing this not to overheat the piece, or the structure may be altered. It is then rubbed on emery papers of successively finer grades, starting with a fairly coarse one and finishing with one so fine that it looks like glazed paper. This is generally sufficient, but if necessary, the piece is further polished using rouge. In an industrial laboratory, with many specimens to examine each day, these processes are mechanised, to remove some of the drudgery.

The polished surface of the specimen has now to be etched, that is, treated with some chemical agent—nitric acid is often used—which

attacks the various constituents to a different degree. A few seconds'
treatment only are generally needed, and to the naked eye the
polished surface appears hardly any different. But in fact the reagent
has eaten some parts of the metal away more than others, thus the
iron will generally be attacked more, and thus etched deeper, than
the carbon, which is another important constituent of steel. The
matter is not *quite* as simple as that, but this gives an idea of the
principle involved. If the specimen is now examined with a microscope, using suitable lighting to accentuate the differences of level,
the size and disposition of the individual crystals can easily be seen.

In principle, two kinds of illumination are suitable for the purpose.
Strong oblique lighting, that is, a beam of light directed from one side
of the instrument almost parallel to the surface of the specimen, shows
up differences in level in a most striking fashion. The reader can
demonstrate this for himself, without a microscope, simply by holding a page of this book so that the light from a desk lamp falls upon
it obliquely. When the correct angle of illumination is found, the
minute indentations in the paper caused by the impression of the
printing types will be clearly seen, though they are normally quite
invisible. Oblique lighting, however, suffers from two disadvantages.
As will be obvious from the experiment just described, its effective-
ness depends upon the exaggeration of minute differences in height
due to the disproportionately long shadows which they cast. In the
microscope, this may easily give rise to a misleading image. The
second disadvantage is that oblique lighting is very difficult to arrange when a high-power objective is being used, since such objectives have to work very close to the surface of the specimen.

It is therefore much more usual to work with vertical illumination,
and a metallurgical microscope is specially designed to provide this,
as well as to facilitate the examination of metal specimens generally.
Since all the specimens to be examined will be opaque, the substage
condenser and its accessories can be dispensed with, and the stage of
a metallurgical microscope is made solid, without a hole in its centre
such as is found in most microscopes. The essential parts of the
illuminating system are sketched in Fig 23. It will be seen that light
enters the side of the microscope tube and is reflected vertically
downwards through the objective by a reflector which is usually a
very thin and perfectly flat piece of glass. Thus, in this arrangement,

the objective itself acts as part of the condensing lens system. The light reflected from the specimen then passes back through the objective—and through the reflector—to form the final image in the usual way.

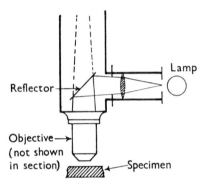

Fig 23   The principal parts of the 'vertical illuminator' of a metallurgical microscope. The fine full lines show the path of the light rays from the lamp to the objective, and thence to the specimen; the broken lines show the light path of the rays from the specimen towards the eyepiece (compare Fig 21)

Obviously, neither reflection nor transmission can be entirely perfect, and some losses are inevitable, but careful design and construction keeps them to the minimum, and the system is very successful in practice. Another method has been devised, and finds some use, in which the transparent reflector is replaced by a mirror with a hole in its centre. This reflects the light through a system of lenses arranged concentrically around the outside of the objective, which thus plays no part in the illumination of the specimen. With this method, there is no interference with either reflection or transmission of light, but this advantage is gained at the expense of greater complication, and therefore of increased cost.

Wherever possible, metallurgical microscopes make use of a 'built-in' source of light, but for some purposes this is inadequate, and an outside source must be used. Where this is the case, it is clearly undesirable to move the microscope tube to any great extent for focusing purposes, and a metallurgical microscope is arranged so that the stage, rather than the microscope tube, can be moved for this purpose when necessary. This feature is also useful when changing specimens.

By far the commonest structural metal, and therefore the one most often examined microscopically, is steel. However, the metallurgy of steel is somewhat complex, and the principles involved will here be illustrated by an example of type metal. This is an alloy of lead, tin, and antimony, and, as the name suggests, it is the metal from which printing type is cast. Very similar alloys are used to line bearings, and probably this application is more important from the engineer's point of view.

A general impression of the appearance of the polished and etched surface of type metal, magnified about 150 times, is shown in Fig 24.

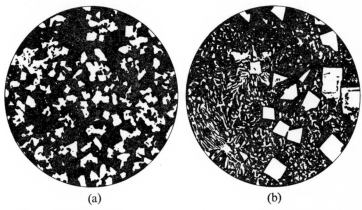

(a)   (b)

Fig 24   Drawings showing the appearance, ×150, of the polished and etched surfaces of specimens of type metal (a) rapidly cooled from the molten state; (b) slowly cooled

The crystalline structure is at once apparent. The large, more or less square white patches are actually cubic crystals containing approximately 50 per cent tin and 50 per cent antimony. The bulk of the lead in the alloy, together with much smaller proportions of tin and antimony, forms the background mass, the crystal structure of which, though it exists, is much less obvious, and in this drawing it appears as a uniform black ground.

This figure also demonstrates how different physical treatments can affect the crystal structure of a metal or alloy. The composition

193

of the alloy in each part of the illustration is the same; chemical analysis would reveal no difference between them. The samples might, indeed, be taken from the same batch of molten metal. However, the specimen in (a) was quickly cooled from the molten condition; the tin-antimony crystals did not have much time in which to form, and so remain small and evenly distributed. Specimen (b) on the other hand was slowly cooled; the tin-antimony alloy, having a rather higher melting point than the remainder of the metal, formed crystals as before, but now these crystals had time to grow to a large size before the whole of the metal solidified. Moreover, since the alloy of which they are composed is lighter than the remaining metal, the crystals tended to rise and congregate in the upper part of the specimen (which actually appears on the right-hand side of the drawing).

These crystals are also much harder than the bulk of the metal, and they can be made still harder by adding other metals, such as copper, to the alloy. The pictures make it obvious why such metals are used for bearings: they consist of numerous crystals of hard metal, well adapted to resist wear, embedded in a 'matrix' of much softer metal, which can to some extent deform so as to adapt itself to the contours of the shaft which runs in the bearing. Thus a well-fitting and wear-resisting bearing is easily obtained. The value of laboratory examination is here well shown. If one were attempting to develop a new bearing metal, specimens of the various experimental mixtures would be examined microscopically, and only those which exhibited the appropriate structure would be selected for further practical testing. It is easy to see how much time and trouble would thus be saved.

STRESS AND STRAIN

There are a great many parameters (see page 95) and mathematical relationships which may at some time be encountered in the study of strength of materials, but fortunately, for the purposes of this book, it is not necessary to consider most of them. There are, however, three important concepts to which the reader should be introduced: these are stress, strain, and the modulus of elasticity.

Now, it is a matter of common experience that if a force is applied to some object which is fixed so that it cannot move, that object will be deformed. Thus, if you place an indiarubber on the table and press on it with your finger, a depression forms under the finger and the rubber bulges out at the sides. Much the same thing happens with a piece of steel, but in that case one could not see the deformation. It is there, however, and can be measured with sensitive instruments. The force applied to an object, whether by pressing, as in this example, by pulling on it or in any other way, is called *stress*. Since such a force is necessarily applied at the surface of the object, it can be measured in terms of pressure, and this is still valid even if the stress is a tensile one, tending to pull the object apart, or a shearing one, tending to displace one part of it relative to another, as by twisting. Stress is therefore measured in pounds per square inch (lb/in^2) or kilograms per square centimetre (kg/cm^2). We shall stick to the former in this chapter.

The effect produced upon an object by an applied stress is called *strain*. It can be recorded in several different ways, depending upon the nature of the stress. In our example of the indiarubber we might, for instance, record it as the percentage change in thickness under the stress. If the experiment with the rubber were repeated with a piece of Plasticine, the result would be much the same, but with one important difference: whereas the rubber would return to its original shape after the stress was removed, the Plasticine would not do so. In fact, for any given material, there is a definite limit to the stress which may be applied, if the specimen is to recover after it is removed. Beyond that limit, the deformation will be permanent.

So far as structural materials are concerned, we are only interested in the behaviour of materials within the *elastic limit*. A structural member or machine part which deformed permanently under load would be useless. This is fortunate for us, for the behaviour of materials outside the elastic limit is much more complicated, though an understanding of it is necessary, for example, if we want to study the processes of metal forming, such as stamping, rolling, or wire-drawing. It will not be considered in this book, however.

Within the range specified, the behaviour of materials is described by Hooke's Law. This was named after its discoverer, Robert Hooke (1635–1703), who studied the behaviour of spiral springs. These represent one of the best and easiest studied examples of elastic behaviour, but in fact all materials behave more or less the same, irrespective of form. Hooke's Law states that 'Strain is proportional to stress', or, to state it mathematically for the simplest case:

$$p = Y\frac{l}{L}$$

In this equation, p represents the applied stress, L is the original length of the specimen, l is the change in length due to the stress, therefore (see above) l/L is the strain. This leaves us with a factor Y. It is found by experiment that this factor is always the same for a given material. It must therefore represent some fundamental physical property of the material, and it is called the *modulus of elasticity*, or Young's modulus, after the physicist, Thomas Young (1773–1829). The importance of this quantity will emerge when we have seen how the various aspects of the strength of materials are measured, which we shall now do.

METHODS OF TESTING MATERIALS

No modern engineering designer is content to leave the selection of his materials to chance. At the very least, he will specify definite named or numbered materials, the properties of which are guaranteed by the makers. If he is in any doubt, there are laboratories which will test materials to see whether they are suitable for his purpose. Large firms go further than this, and maintain their own test houses. This attitude is now spreading into other industries, and there are a great many laboratories up and down the country fully engaged in the testing of materials.

Tensile testing

Tensile testing is perhaps the commonest form of strength testing. Not only are materials often used in tension, but it is known from experiment that several other properties are more or less closely correlated with the tensile strength, so that this one measurement can tell us a great deal about the material. The principle of tensile testing could not be simpler: we apply an end load and watch what happens. In practice, however, it is a little more elaborate than that.

A typical testing machine is shown on page 203. It is called a 'universal testing machine', because it can be set up to perform several kinds of test, and tensile, compression, and shear tests can all be made on this machine. To make these tests we need, basically, three things: a means of applying a load to the specimen, a means of measuring the applied stress, and a means of measuring the strain produced. In a very small machine the load could be applied directly by means of weights, but in larger machines it is usually applied either hydraulically or mechanically, an electric motor providing the driving force in the latter case. The load is applied at the lower end of the specimen in the machine shown. Connected to the upper end is the apparatus for measuring stress. This is the weighing system, which is in effect, simply a large scale, working on the principle of the lever balance described in Chapter 3, the levers being arranged, as a rule, to give a ratio of 2000:1, that is, a weight of 1lb on the weighing arm represents an applied load of 2,000lb.

With the machine set for tensile testing, the specimen is held firmly by its ends in the clamps seen at about the centre of the machine. In order that the results of tests on various machines and in different laboratories shall be comparable, these specimens are carefully prepared and machined to standard dimensions. Among other things, the cross-sectional area is standardised, so knowing this, and being able to measure the applied load, we can deduce the stress in pounds per square inch. In fact, the machine is usually arranged to indicate this directly, either on a suitable dial, or, more conveniently, as a permanent record in the form of a pen trace on paper.

All we need now is the method of measuring strain. This is not an integral part of the machine, as it varies according to the kind of test being carried out. For the tensile test it is an *extensometer*, which is an instrument capable of measuring the increase in length of the specimen under the influence of the applied load. There are several kinds of extensometer, but basically they are all very sensitive length-measuring appliances; an accuracy of $\pm 0\cdot 0001$in is required. The extensometer is attached to the specimen.

To make a tensile test, the specimen is clamped as already described, and the extensometer is attached. The load is then applied. The speed of loading is important in some cases, and is always recorded in the report of the test. As the load is applied, the extensometer indicates a gradual increase in length, and this is noted, as is the stress, if the machine is not equipped to record this automatically. The increase in length takes place in step with the increasing stress, up to a certain point, at which the rate of increase in length, that is, strain, rises faster than the rate of increase of stress. This point is the *limit of proportionality*, and corresponds to the elastic limit mentioned earlier. In most materials this increase of strain over stress is fairly gradual, but in some, of which iron and soft steel are important examples, it is quite rapid and sudden, and is termed the *yield point*. The reading of stress at this point tells us the maximum stress to which structures made of the material in question could usefully be subjected in practice, but it is usual to carry the test beyond this point to ascertain the maximum to which it could possibly be subjected. The extensometer is removed, and loading is continued. Strain now increases much more rapidly than stress, until there is, usually, a sudden 'necking down' of the specimen, and it breaks. The stress at this point is the *ultimate tensile stress*, and the figure for this is the one usually quoted when comparing the strength of various materials.

Compression and shear tests

Testing of compressive and shear strength is done on the same machine as is used for tensile tests, only the machine set-up being altered, which is a great convenience. To make a compression test, the direction of application of the load is reversed, and instead of being pulled apart, the specimen, which is again machined to standard dimensions, is squeezed between two *platens*, which are flat plates, the upper one being connected to the weighing mechanism as before. In a compression test, there is no total rupture, such as occurs in a tensile test, but the specimen deforms more or less abruptly at the yield point. This in fact gives us all the information we require, for a structural member in compression has, for all practical purposes, failed if it deforms under load, whether or not it actually breaks.

In a shear test, the same arrangement of the machine is used, but the specimen, which is generally in the form of a flat plate, is placed between a *punch* and a *die*. The die is simply a piece of hardened steel with a hole in it, and the punch is another piece of hardened steel which exactly fits the hole. The dimensions of both are standardised. They are fixed to the platens of the machine so that the punch is aligned with the hole in the die. Load is applied until the specimen fails, a piece being cut out of it and driven into the die by the punch.

The stress at this point is the shear stress of the material. Actually, only the initial failure is due to shear, a rather different mechanism being responsible for the complete cutting through of the part, but again, it is the point of first failure in which we are interested for practical purposes.

Impact testing

It is a matter of common experience that many materials—glass is probably the best known—while being relatively, or even extremely strong under constant load, will break quite easily if subjected to only a slight shock. This has always been recognised, but some years ago attention was attracted by a series of failures in service of structures made of materials which had hitherto been regarded as perfectly adequate to their duty. It was shown that these failures took place as the result of shock loading, and the term 'brittle failure' was coined to denote this phenomenon. One reason for this sudden increase in brittle failure was the increased use of welding, a process which can considerably modify the properties of the metals on which it is used. More will be said about this in the next chapter.

Thus, we need tests to distinguish between materials which, although they may have equivalent tensile strengths, differ in their resistance to shock loading, or, in plain terms, which are tough and which are brittle. It is also important to be able to detect any tendency to brittle failure. Finally, not only the composition of a material, but the treatment it receives, can have a profound effect, as can easily be demonstrated by taking a couple of ordinary sewing needles and making them both red-hot. Allow one to cool down as slowly as possible, by slowly reducing the heating flame or (the traditional method) by allowing it to cool down in the fire overnight, and cool the other rapidly by plunging it into cold water. If an attempt is now made to bend each of them, it will be found that the quickly-cooled one is very hard, and difficult to bend, but when enough force is applied, it will break. The other needle is much softer, and will bend quite easily, but does not break. Heat treatment of this kind is very common in engineering, and its effects must be measured.

For these purposes a group of tests called 'impact tests' has been devised, the two commonest being the Izod and Charpy tests. The same machines are used for both, the principal difference being in the form of the test-piece. An impact testing machine is basically a very simple piece of apparatus, and consists essentially of a swinging hammer, or pendulum, of known weight, which can be raised to a fixed height, and then released, when it swings down to strike and break the test piece held in its path. It is found that the height to which the pendulum swings after breaking the specimen, is proportional to the strength of the specimen (this is fairly obvious: the stronger the specimen the more force will be expended in breaking it), and this height is recorded on a simple dial.

For the Izod test, the test-piece is 10mm square and 75mm long, and a notch, of triangular form and standard dimensions, is machined 28mm from one end. The piece is held in a vice, this end being uppermost, and the notch level with the vice jaws. The pendulum is allowed to strike the test-piece as described above, the point of impact also being laid down in the conditions for the test, and the distance which it swings after the piece breaks, which it will do at the notch, of course, is the measure of the Izod strength of the material.

In the Charpy test, the test-piece measures 10mm square and 55mm long, the notch in this case being in the middle. A number of alternative forms is laid down for the notch, according to the exact nature of the test to be made. In every case, however, the specimen is supported horizontally in the machine and held at its ends, and the pendulum strikes it in the centre. Again, it breaks at the notch, and the distance travelled by the pendulum afterwards is the measure of the Charpy impact strength.

Creep and fatigue testing

Creep testing is basically no different from tensile testing, but the specimen is not loaded to the yield point. Instead, a predetermined stress is applied and maintained for a definite length of time. Extremely sensitive extensometers enable the extent of any creep which takes place to be determined within a much shorter period than would be required for it to show itself under working conditions. The actual conditions of the test may be rather more complicated, because it is often necessary to determine creep at high temperatures, and both the machines and the measuring instruments must then be capable of operating at these temperatures.

Reference back to the description of the phenomenon of fatigue, on page 189, will show the kind of test needed to evaluate it. There must be some means of applying a load, which may, and in most cases will, be much less than the breaking load, many times in succession. Possibly the most spectacular example of a fatigue test was that devised in connection with the 'Comet' aircraft. This machine had been, to all appearances, most thoroughly tested by its builders, but soon after production models had entered airline service, there were a number of disastrous, and apparently inexplicable crashes. Intensive investigations eventually led to the belief that these were due to fatigue failures in the pressure cabins. In a modern high-altitude aircraft, the atmosphere inside the passenger cabin is kept at a pressure only a little below that at normal sea level, whereas the air outside, when the machine is at its cruising altitude, is far below that pressure. The cabin is therefore subjected to stress while the aircraft is in flight, and this stress is applied and removed each time the aircraft takes off and lands. Thus we have the conditions in which fatigue failure could arise.

To test this theory, a huge tank, large enough to contain an entire aircraft, was built, and in this an actual machine was subjected to alternating pressures, applied hydraulically, to simulate the pressurisation and depressurisation experienced in practice. Sure enough, after a number of cycles approximately corresponding to the length of time the crashed machines had been in service, the cabin failed. It will be appreciated that the test was 'speeded up' to take days instead of months, and this is one of the most valuable features of this kind of testing.

To reassure any reader who may have occasion to fly by 'Comet', it should be said that the aircraft was re-designed, in the light of the test findings, and is now one of the safest, because one of the most tested, aircraft which has ever been in service.

Of course, it is not normally a practical or economic proposition to test an entire aircraft, or other structure, to destruction, and it is more usual to test samples of material for fatigue resistance. The necessary alternating load can be applied in a number of ways. One simple and obvious method is by means of a crank and lever, which is connected to the test specimen either directly or via a spring. It can also be done electromagnetically, which has the advantage that the reversals can take place very rapidly indeed, enabling years of service to be simulated in a few hours' testing. If high loadings are required, they are applied hydraulically.

Probably the commonest methods, however, are either to use out-of-balance centrifugal forces, or to attach a dead weight to the end of the specimen, which in this case is in the form of a long rod. This is then rotated on its own axis, and the effect is to produce a bending force, which is reversed twice in each revolution. However the stress is applied, some means of counting the number of cycles before failure takes place, is incorporated in the testing machine, and the number of cycles to failure is the measure of the fatigue strength of the material. It is a peculiarity of the phenomenon of fatigue that no two tests give exactly the same result, although the results are statistically the same (see Chapter 2), for a curve of applied stress plotted against the number of cycles to failure has the same shape for a given material, independent of the actual stresses or numbers of cycles.

This kind of behaviour is one of the things which makes it difficult to propound a satisfactory theory of fatigue. It is fairly certain that changes in the internal structure of the material are involved, but it has also been shown that poor surface finish, cracks, and inclusions of foreign matter can all increase susceptibility to fatigue failure. Moreover, the two latter may only be of the most microscopic nature. It is thus important to discover such imperfections before a part is placed in service, and more will be said about this in the next chapter.

HARDNESS TESTS

The last property of materials with which we shall here be concerned is hardness. Some materials are naturally harder than others, of course, and it may be necessary to make comparative tests, but, as has already been pointed out, the hardness of most metals can be profoundly affected by heat treatment, and a great deal of engineering practice is concerned with altering the properties of materials in this way. Other treatments to which metals may be subjected in the workshop, such as hammering, rolling, pressing, etc—what is generally referred to as 'cold working'—can also affect their hardness. Thus, the primary uses of hardness tests are to ascertain whether some treatment has been correctly carried out, and whether some component

Page 203 A Denison model T42B4 universal testing machine, with recorder and strain rate indicator, as installed at and with the kind permission of Stewarts and Lloyds Limited, Corby. The specimen being tested can be seen between the platens on the left

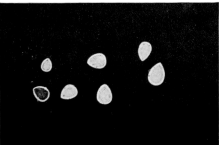

Page 204 (*above*) Photostress analysis: the pattern of stresses in an aircraft landing-wheel revealed by a photoelastic coating; (*below*) an immersion contrast photograph of one real and six synthetic rubies

has the right hardness for the service required. In this connection it should, perhaps, be pointed out that it is quite possible for a part to be too hard as well as not hard enough. Hardness testing is therefore quite frequently carried out as a routine measure on finished parts, and may be regarded as a method of non-destructive testing, which is dealt with in the next chapter. It is usual, however, to consider the subject with the testing of materials, and this convention will be followed here.

The first practical measure of hardness was Mohs' scale, named after its inventor Friedrich Mohs (1773–1839), which was introduced to measure the hardness of minerals, for which purpose it is still extensively used. It is an arbitrary scale, and in some ways rather crude, but it has the great merit of being easily applied, and needs no elaborate apparatus. Mohs took ten different minerals, and arranged and numbered them in such a way that each would scratch the one above it on the scale, but not the one below it. Thus a mineral of unknown hardness can be assigned to an approximate place in the scale by seeing which of these minerals it will scratch, and which it is itself scratched by. The scale is:

1 talc
2 gypsum
3 calcite
4 fluorite
5 apatite
6 orthoclase
7 quartz
8 topaz
9 corundum
10 diamond

An experienced mineralogist can usually make a fairly accurate estimate of the Mohs' hardness of a mineral by seeing how easy or difficult it is to scratch it with something of known hardness, and this fact points the way to the general principles of most other hardness tests.

There are several tests of hardness in use today, each requiring its own special machine. Strictly speaking, the scales are arbitrary, and apply only to the machine being used, but it is found in practice that the results obtained by various tests are comparable, and conversion tables for the various systems have been drawn up. This proves, incidentally, that hardness is a real property of the material, and is independent of the method used to test it.

One of the most generally useful tests, though not one particularly suited to routine inspection procedures, is the Brinell hardness test. In this, a hardened steel ball is forced into the specimen under test, by a known load. The load is applied mechanically or hydraulically, and weighed, the principle being exactly as in the testing machines described earlier. The ball will make an impression in the part being tested, and, fairly obviously, the softer the part, the deeper will be the impression for a given load. The diameter of the impression, which is proportional to its depth, is measured with a microscope having a graduated scale in the eyepiece, and this measurement, in conjunction with the applied load, gives the Brinell hardness of the specimen, which is expressed as a number.

The factors of the hardness number are thus load and depth of penetration, and since load can easily be standardised, it ought to be possible to measure depth of penetration to get a direct reading of hardness. This in fact is done in the Rockwell hardness-testing machine, which uses either a steel ball or a diamond to indent the specimen, the hardness being read off directly on a dial. This machine is suitable for use in routine inspection procedures. Once it has been adjusted to accept a certain component, no particular skill or knowledge are required from the operator. The part to be tested is placed on a platform, a small lever is operated to locate the indenting point and apply the load, the hardness is read on the dial, the lever released, the part is accepted or rejected as the case may be, and the next component is tested in turn.

A very similar machine, but with a different scale, is the Vickers Diamond Pyramid Tester, the nature of which can well be gathered from the name.

All the machines so far described do damage the surface of the specimen, very slightly, it is true, but noticeably. Usually it is possible so to arrange matters that a finished component is tested in some inconspicuous place, but this is not always convenient. For finished parts which must not be marked or damaged in any way, the Shore Scleroscope is useful. This is a very ingenious device. If you hit a piece of lead with a hammer, the blow feels 'dead', but if you strike a blow of equivalent weight on a piece of steel, the effect is quite different; the hammer tends to rebound. The Shore Scleroscope uses this principle, in a rather more refined form. A diamond-tipped hammer is allowed to fall on the specimen to be tested; the height through which it falls is standardised, and the release mechanism must be carefully designed so that there is no possibility of the hammer being 'projected' at the specimen. It must fall entirely under the influence of gravity, or the results will not be repeatable. After striking the specimen, the hammer will rebound, and the extent to which it rebounds is a measure of the hardness of the specimen.

8

... But Don't Break it

In the last chapter we looked at some ways of testing materials and component parts. Most of these involved some damage to, or even the complete destruction of the specimen. Consequently the tests are usually confined to a small sample taken on the assumption that the sample is representative of the bulk, and for many purposes this is adequate. Statistically, we can predict the percentage of failures which will occur as a result of testing only samples, and the size of the samples is adjusted to keep this percentage within a predetermined figure. This is primarily a matter of economics: the price of a motor car would be very much higher if it were demanded that every part put into every car should be tested. Again, it may simply not be practical: it is very annoying when an electric bulb burns out after only a few hours' use, but to test all of them completely would result in there being none for sale!

However, there are occasions when these arguments cannot be accepted. A statistically certain failure in a motor car, if it is infrequent enough, is tolerable, but in an aeroplane or a space vehicle it is not. Every part, even the nuts and bolts, must be tested, yet the tests must not themselves damage or weaken the parts or lessen their life expectancy.

Again, the behaviour of a complicated structure may be very different from that of its parts. To be certain that it will behave as intended, the entire structure should be tested. But modern technology has given rise to very large and complex assemblies, very expensive in materials and labour, and it is out of the question to sacrifice even one of them for testing to destruction—in many cases the job will be only 'one off' anyway. Welding is often extensively used in such structures, and it is obviously impossible to open every weld for examination, and a random sampling, owing to the 'human element' present in the welding operation, is not applicable—operator fatigue or similar factors may result in a 'bunching' of faults at one time, while at another, when everything is going well, there may be a long run of perfect welds.

The technique of examining materials and structures without destroying or damaging them, or altering their properties, is called *non-destructive testing*, often abbreviated simply to NDT. The ideal is an old one, and an early example will be mentioned in the next chapter, but as a technical and scientific procedure, it is a modern development, brought into being by the considerations outlined in the last paragraph. A whole armoury of tests has been developed, and nowadays it is possible to find out quite a lot about a part or an assembly, sometimes without even taking it out of service. We shall now look at some of these tests, starting with the simplest, and progressing to some of the later advanced developments.

VISUAL EXAMINATION

The simplest and most obvious way to find out about something is simply to look at it, and an informed and careful scrutiny is in fact the oldest method of non-destructive testing, and by no means superseded today. We shall have occasion to return to this theme in another connection, but for the moment let us confine our observations to the field of engineering. Every manufactured part is subjected to some degree of inspection, even if this only amounts to an examination by the operator or craftsman who made it. Frequently, of course, and invariably in quantity production, this process is much more rigorous and formalised, and is the responsibility of specialist staff. The primary object of inspection is more often than not to ensure that the article has been made to the correct dimensions, but this aspect of the matter has already been dealt with in Chapter 3. Often, however, the inspection will be concerned with the less definable, but none the less real, questions of 'quality' and 'finish'. Careful visual examination will disclose much to the expert, but much more can be learnt if unaided eyesight can be supplemented at need by suitable optical aids.

The most obvious example of this is the use of a magnifier, and the well-equipped inspector will always have an adequate selection of these at hand. From the use of the simple magnifier, it is but a step to the microscope proper, and this, described more fully in Chapter 6, is a most important instrument of non-destructive testing.

Examination of surface finish comes to mind at once in this connection, and the microscope is indeed very often used for this purpose, though nowadays it is supplemented and to some extent replaced by the more sophisticated methods described a little farther on. However, in industrial practice it would be rather inconvenient, and extremely fatiguing, for an inspector to spend the greater part of his day looking through a microscope of the conventional design described earlier (though there are a few applications where this is necessary). Instead, matters are so arranged that a magnified image of the object under examination is projected on to a screen, where it can be viewed in comfort.

This method has another advantage: it is a very simple matter to mark the screen with dimension lines, so that the dimensions of an article and its surface finish can be checked in a single operation. The entire output of certain small components is sometimes inspected in this way, in what amounts to a 'mass-production' operation. Small articles whose geometrical form is important are nearly always checked by this method, the usual procedure being for the drawing office to prepare an enlarged drawing of the article to the appropriate scale. This drawing is mounted on the screen, so that the projected image of the actual part under test coincides with it. Any deviations from the acceptable limits of size, form, or finish can then be detected at a glance. In engineering, screw threads and the teeth of gear wheels are examples of objects for which this kind of inspection is very suitable.

However, it may well happen that the outside appearance of an object is not the only, perhaps not even the important, aspect of it which must be examined. Thus, it is necessary to examine castings, not only for surface finish, but to make sure there are no small cracks which could cause trouble later. But many casings are hollow—'cored out' as the engineer expresses it—and the internal shape may well be very complicated, and defy direct visual inspection. The same thing applies to many welded assemblies.

In these instances, the inspection mirror, that well-known symbol of the dentist, is a valuable ally. These are made in a variety of sizes and shapes, and can be introduced into the most awkward openings and corners. They can also be arranged to provide some degree of magnification, which is often most useful. These simple gadgets have been developed into a whole range of inspection instruments, in which lenses, prisms, and mirrors are combined so as to make possible the examination of all sorts of inaccessible features. One example of this, and a striking illustration of testing which simply must be non-destructive, is the examination of the various cavities of the human body. Not only the nose and mouth, but the ears, lungs, and practically the whole of the digestive and genito-urinary tracts can be probed by instruments of this kind, which are usually arranged to transmit the light needed to make the examination as well. Some of them indeed carry their own light sources in the form of tiny, but powerful, bulbs specially developed for the purpose.

In the engineering field, these instruments are exemplified by the Borescope, which makes it possible to examine the insides of tubes and bores as small as 0·02in diameter. A good example of its application is in the inspection of gun barrels, the surface finish of which must be beyond reproach if good results are to be had.

Still considering surface finish, many photographers must have had the experience of photographing a surface or background which they believed to be perfectly smooth, only to find, when the final print was prepared, that it was anything but. This ability of the camera to exaggerate differences of contour and texture hardly noticeable to the eye can be a valuable ally in the examination of surface finish, and it is often a matter of routine to photograph the object to be inspected. Suitable lighting for the photography, together perhaps with a judicious choice of film and filter, and some degree of enlargement of the final print, can make comparatively easy what would, to the unaided eye alone, be a very exacting task.

While on the subject of photography, it may be recalled that many amateur photographers, whose interest is in colour slides, have been troubled by 'Newton's rings'. These occur when two surfaces are nearly, but not quite, in contact—when they are separated by a space of the order of 0·0001in. They are the result of a phenomenon termed 'the interference of light'. If we have a suitable reference surface which is perfectly flat, within a few millionths of an inch, we can use this phenomenon to investigate surface finish and flatness. The reference surface is made of glass or quartz, worked to the degree of flatness mentioned, and is termed an 'optical flat'. The surface to be tested is cleaned, and the optical flat is placed in contact with it. If the two surfaces were entirely in contact—which is so unlikely that it may safely be characterised as impossible—nothing would be seen. The same result, however, would be obtained if they were out of contact by more than about 0·001in, so the surface to be examined must be fairly flat, relatively speaking, for this test to have any meaning Granted that it is, Newton's rings will appear, and their number and shape form a very sensitive indication of the condition of the surface. In effect, this method magnifies the irregularities by something like two thousand times.

The methods so far described represent, so to speak, extensions of the visual sense. But there is another very common method of evaluating surface finish: we can feel the surface. This method has been used in the workshop since time immemorial, but it does not lend itself particularly well to the purposes of quantitative inspection. However, instruments have been invented which, in effect, extend the sense of touch. They are called profile recorders, and work by moving a fine stylus over the surface to be examined. The movements of this, due to the surface irregularities, are magnified mechanically or optically, and the result is presented as a pen tracing or a series of readings. Profile recorders are very sensitive—the standard unit of surface finish used with them is the micro-inch, which is one millionth of an inch.

PRESSURE AND LEAK TESTS

Since the early days of the Industrial Revolution, 'steam' has always been to some extent a synonym for 'power', and evidence is not lacking that the early engineers were well aware of the magnitude of the forces which they were trying to control. James Watt would never countenance pressures of more than a few pounds per square inch for his engines. It was realised almost from the first, however, that much higher pressures than this were needed for anything like real efficiency, and attempts were soon made to work engines at higher pressures. The almost immediate outcome of these attempts was boiler explosions, and engineers soon recognised that for their own safety, if not for that of the public at large, some way of testing a boiler before putting it into service was essential.

Pressure and leak tests have therefore been a feature of engineering practice since the earliest times, and they are, of necessity, non-destructive. Regular inspection and testing of boilers and other pressure vessels has, in fact, been a legal requirement for many years. The first part of such an inspection consists of a careful visual examination, to make sure that joints are properly made, that the thickness of plates is sufficient, and that the general design of the vessel is sound. Then the vessel is subjected to a pressure considerably above that at which it is intended to work. Double the working pressure is often specified. Of course, such a pressure cannot be produced in a boiler by raising steam in it, for if there should be a fault the resulting explosion would be even more disastrous than one under working conditions; this consideration also applies to vessels intended to contain air or other gases under pressure.

The pressure is therefore applied *hydrostatically*. This means that the vessel is completely filled with a liquid, generally water, all the openings except two being sealed. To one of these openings a pressure gauge is connected, and to the other a force pump. The test pressure is applied by forcing in more liquid by means of the pump; such a pump has a very high mechanical advantage, and high pressures can be attained quite easily. Liquids are virtually incompressible, and should there be a fault which causes the vessel to fail under test, the pressure is released automatically and immediately, and there is no danger of explosion.

A hydrostatic test proves the soundness of the structure of the vessel, and it will also reveal any gross leakages, but leaks are sometimes very small and difficult to 'pinpoint'. Where complete freedom from leakage is important, other tests must be applied. The basis of these is simply the same as the well-known method of testing a bicycle tire for punctures, and if the vessel to be tested is small enough, the procedure can be exactly the same. Compressed air is pumped into the vessel, which is then completely immersed in a tank of water. A stream of bubbles will reveal any leakage. If the vessel is too large for this, it may be charged with compressed air as before, and the outside is then coated with soap solution. A leak will produce a bubble in this. This is a good test for general purposes, but very tiny leaks, such as might be caused by porosity in a casting, are not easy to detect in this way, and where they would be important, a more sophisticated method has to be used.

This is extremely ingenious, and depends upon one of the facts of chemistry which were discussed in Chapter 4. The outside of the vessel to be tested is coated with phenolphthalein. This is a colourless liquid in normal circumstances, but it is an indicator, and is in fact so used in the laboratory. In presence of alkalies, it turns bright red, the reaction being a very sensitive one. Instead of compressed air, the vessel is filled with ammonia. Ammonia is a gas at normal temperatures and pressures; the liquid sold as 'ammonia' in hardware shops is a strong solution of it in water, and anybody who has had occasion to use it knows that it is strongly alkaline. So with the set-up described, any leak, even a very tiny one, is rapidly shown by a red stain on the outside of the vessel.

PENETRANT TESTING

It was mentioned above that porosity in castings may be a source of trouble, but it is extremely difficult to detect by mere visual examination. A rather similar fault, also found particularly in castings, and equally difficult to see in the ordinary way, is the presence of very tiny cracks. These, if undetected, could easily be the starting points of fatigue failures in service, so methods of revealing these flaws have always received great attention, and in fact, such a method was one of the earliest forms of what we should nowadays call non-destructive testing. This was the 'oil-and-whitewash' test. To apply it, the casting (the process is not confined to castings, but these were by far the most commonly tested, so we will continue to refer to them for simplicity) was first completely covered—immersed, if possible—in a fairly thin oil. The surface was then carefully and thoroughly cleaned. The oil would naturally have penetrated into any flaws there might be, and would remain in them after the surface was cleaned. The whole casting was now coated with ordinary whitewash, and left until this had completely dried. After a time, the presence of flaws would be revealed by oil stains on the whitewash, caused by seepage, aided by capillary effects, the whitewash acting in a manner not unlike blotting paper. Anyone who has had the misfortune to drop a tiny spot of oil on a distempered wall knows how it spreads; in this way the size of the cracks is, in effect, magnified.

Sometimes, to make the oil seepage easier to see, dyes were added to the oil. This was the starting point for the development of modern penetrant testing methods. The oil has been superseded by special liquids developed for the purpose. These are obtainable commercially, and are of two principal kinds, dye and fluorescent types. The dye penetrants offer high colour contrasts, so the flaws are easily seen. The fluorescent types appear more or less colourless under ordinary lighting, but if they are illuminated by ultraviolet radiation they fluoresce, that is, they glow strongly. The slightest trace of penetrant is enough to show this effect, thus making the test a very sensitive one, but a darkened room and an ultraviolet lamp are needed. Dye penetrants can be used in ordinary daylight without special equipment.

To make the test, the surface is first thoroughly cleaned. This is essential, and chemical methods of cleaning are generally used, both for speed and efficiency. The penetrant is next applied, and all excess is removed. This is the most critical part of the whole procedure. Nothing may be left on the surface, but at the same time, nothing should be removed from any possible flaws. The removal successfully accomplished, a developer, which is usually talc or some similar substance, is applied to the surface, and after allowing some time for the penetrant to be drawn out of the flaws by capillary attraction, the surface is inspected and the result interpreted. A skilled operator can deduce quite a lot about the nature of the flaws from the appearance of the penetrant stains on the surface.

RADIOGRAPHY

Most practical mechanics have had the experience of having to scrap some piece of work, perhaps a very elaborate one, because of some hidden and unsuspected fault in a casting or piece of material which has not been revealed until a fairly late stage in the machining process. 'Blowholes' in castings are a common cause of this trouble; they are caused by gas bubbles formed in the molten metal, which have been unable to escape before the metal solidified. Many a time the unfortunate in this position must have fervently wished that there was a method of seeing the interior of the casting before starting work. The technique of radiography offers this facility to some extent.

Ordinary light cannot, of course, penetrate metals or many other common structural materials. We have, however, seen, on page 133 that visible light represents only a small fraction of the electromagnetic spectrum, and that other radiations, of shorter wavelength, exist. Some of these can penetrate substances, including metals, which are opaque to ordinary light. The human eye is insensitive to these radiations, but fortunately, photographic plates and films are very sensitive to them. Here then, is a very powerful technique of non-destructive testing.

The form of radiation most commonly used for the purpose is X radiation. Nearly everyone nowadays has some idea of what X-rays are and what they can do, owing to their extensive use in medicine—a perfect example of non-destructive testing, incidentally. X-rays were discovered in 1895 by the German physicist, W. K. Röntgen, and are often, more particularly on the continent, called 'Röntgen rays', in his honour. He called them 'X-rays' because he did not understand their nature, *x* being the conventional symbol for the first unknown quantity in algebra.

Röntgen was experimenting with a Crookes tube, that is, an apparatus in which an electrical discharge takes place between two electrodes in a vacuum; his tube had been modified by the inclusion of a third electrode. The discovery was the consequence of one of those happy accidents in which the history of science abounds: a photographic plate which had been in the same room, well covered and protected from light, was found to have been 'fogged'. Why Röntgen should have decided to develop this plate, which by all rights should have been blank, is something which remains unexplained to this day, but he did develop it, with the result mentioned, and he realised that the fogging must be due to some radiation which had been produced in his experimental tube, and which could penetrate materials opaque to visible light.

The modern form of the X-ray tube is not greatly different in principle from the apparatus employed by Röntgen: a schematic diagram of it is shown in Fig 25. The outer protective envelope may be made of metal or plastics. The tube itself is glass; it is hermetically

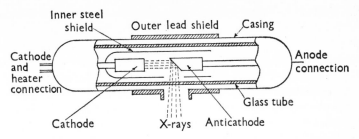

Fig 25   Diagrammatic section through an X-ray tube

sealed, and the interior is an almost perfect vacuum. An electric current at a very high tension, of at least 15,000 volts, and usually much more, is applied between the anode and the cathode. The result of this is that a stream of electrons is emitted from the cathode. We have met this phenomenon before, when examining the electron microscope; it is also the basis of the cathode-ray tube in a television receiver. In the X-ray tube, however, this stream of electrons is caused to strike a third electrode, the target, or *anticathode*. The result of this is the production of X-rays, which leave the tube in a direction determined by the orientation of the anticathode, in the way shown in the diagram. It is usual to surround the tube with a shield of lead, which is fairly opaque to X-rays, leaving only a small 'window' which serves to project the rays in a more or less definite beam. Lead shielding is also used to protect the operators and any passers-by from the effects of the rays, which can be very harmful if one is exposed to them in excess.

In general, X-rays penetrate materials more or less in proportion to their densities. The density of a material is the weight of a given volume of it; it is usually expressed in grams per cubic centimetre. Thus (Chapter 3) water has a density of 1. The density of aluminium is 2·6, whereas that of lead is 11·4; it is the densest of the common metals. Bone is denser than flesh, which is the basis of the medical use of X-rays.

To make a radiograph of, say, a hand, the object is placed in the path of the X-ray beam emerging from the tube window, and a photographic plate or film, suitably enclosed in a container to shield it from ordinary light, is placed behind the object. On development,

217

the film will be found to bear an image of the bones of the hand as fairly clear regions of film, where the X-rays have not been able to penetrate to any great extent, surrounded by a somewhat darker area representing the flesh. Unfortunately, X-rays cannot be focused in the same way as visible light, so all radiographs are in the form of 'shadowgraphs' of this kind, but this is adequate for most purposes. If it is desired to know the *depth* of some object, say the location of a foreign body in medical radiography, it is necessary to make two radiographs, aiming the X-ray beam at different angles for each. The images on the film will then be displaced relative to each other, and the depth of an object can be calculated trigonometrically.

It is possible to use X-rays to examine the interior of metal or other objects in very much the same way. The differences in practice are due to the fact that metals are very much denser than the human body. Thus, if any reasonable thickness of metal is to be examined, much more penetrating rays must be used. We can see intuitively that this means giving the rays more 'kick' by applying more power to the tube, and intuition is correct in this case. More scientifically, the penetrating power of the radiation depends upon its wavelength; the shorter the wavelength the greater the penetration. Wavelength is linked to energy in a very definite way: the relationship can be expressed in a mathematical formula:

$$e = \frac{hc}{\lambda}$$

Here, λ (the Greek letter lambda) is the conventional symbol for wavelength, energy is denoted by e, c is the velocity of the radiation, while the quantity represented by h, which is always the same, is called *Planck's constant*. The exact value of this, or indeed of the other quantities, does not matter for our present purpose; the point brought out by the formula is that shorter wavelengths mean greater energy. The energy of the radiation is derived from the electron beam in the X-ray tube, which in turn is dependent upon the voltage applied. X-ray tubes for industrial use therefore consume a good deal of power.

A second point is that in traversing a considerable thickness of metal, the rays lose much of their energy, and with it their ability to affect a photographic plate, so in order to avoid inconveniently long exposures, an *intensifying screen* is used. This is a plate which is coated with a substance called a *phosphor*, rather like that which forms the screen of a cathode-ray tube. This plate is placed in contact with the film; it glows when the X-rays strike it, and the light thus produced helps to make the exposure. Some of the X-ray energy which would otherwise pass through and be wasted is thus used.

Radiography is extensively used for examining castings and welded joints. Any defects such as porosity, 'blowholes', the inclusion of foreign matter (what welders call 'slag inclusions'), or imperfect fusion of the metal, can be clearly seen on the radiograph. In fact, for important work it is nowadays routine practice to radiograph every weld. The reason why defects show up so clearly is fairly obvious in the light of what has just been said; they differ considerably in density from the bulk of sound metal. The exact position of a flaw can be ascertained by using the 'two-image' technique already mentioned.

It was mentioned above that X-ray tubes for industrial purposes consume a good deal of power. It is also a fact that the tubes themselves are fairly bulky objects, and, since they work at high voltages, further auxiliary equipment in the form of transformers and control equipment is also needed. None of this is particularly disadvantageous in a factory, and indeed it is not very serious anywhere in a highly industrialised country. Many welded structures in particular, however, are not erected under such ideal conditions. Oil and gas pipelines, which may have to traverse hundreds of miles of desert or otherwise difficult and uninhabited country, are a good example. Under these conditions, where every additional kilowatt of power has to be generated on site, using fuel brought over long distances, and every piece of equipment has also to be brought from a distance, and carried with the working party as it moves on, a small, portable source of radiation, using little or no power, would be invaluable.

Such sources are in fact available; they are a by-product of nuclear energy research. It has long been known that one kind of radiation given out by radioactive substances, the gamma-rays (often written γ-rays), are identical in nature with X-rays of very short wavelength. They are thus ideal for the purpose, but until recently natural radioactive substances were prohibitively expensive. With the coming of the atomic energy programme, however, it has become possible to produce radioactive substances ('sources' as they are generally called) artificially, by 'cooking' suitable materials in a nuclear reactor. Thus it is possible to produce a powerful gamma-ray source no more than a few inches long and less than a pound in weight.

This is very suitable for radiography under the conditions just discussed, but it is not without disadvantages of its own, so radioactive sources of this kind are unlikely to supplant X-ray equipment where the latter can conveniently be installed and operated. To start with, although these artificial sources are not nearly so expensive as natural ones would have been, they are not exactly cheap. Further, they do not last indefinitely; the amount of radiation emitted declines with time, and it continues to be emitted whether the source is used or not. The latter fact is the reason for the biggest disadvantage of all; the source cannot possibly be shut off, as an X-ray tube can; it continues to emit radiation all the time, and since this radiation is extremely damaging to living tissue, elaborate precautions have to be taken to keep it where it belongs. This means that the source must be kept in a heavy and bulky lead container, from which it is only removed when it is required for use, and it must be handled and placed in position by means of long-handled tongs or similar devices, which is often rather inconvenient. Sometimes the operator may have to wear special protective clothing as well.

A gamma-ray source has the advantage that it can be placed in confined spaces, for example, inside a pipe, where an X-ray tube could not be taken. The technique of radiography with gamma-rays, apart from the difference of source and the methods of handling it, are just the same as with X-rays.

There is, however, another form of radiography, also developed as a result of nuclear research, which requires different techniques. This is neutron radiography. Neutrons are tiny particles, much smaller than atoms, but larger than the electrons which we have already met. Unlike the latter, however, and in fact unlike most of the particles studied by nuclear science, they have no electrical charge. This is a useful property, for it means that they are not deflected by the intense electrical fields which exist in all matter on the sub-atomic scale, and thus have great penetrative power. They are therefore ideal for the non-destructive examination of considerable thicknesses of dense materials. Unfortunately, they do not affect the photographic plate, so they have to be detected by a technique of 'counting', with an instrument which is sensitive to them. Thus a neutron radiograph cannot be had in the form of a convenient 'picture'; it is a graph, or a series of readings, which must be interpreted, a task calling for some skill. However, there are situations where it is worth while to go to the trouble, and neutron radiography, although a comparatively new development, already finds a good deal of use.

ULTRASONIC TESTING

At one time, not so long ago, a familiar figure on the railways of Britain was the wheel-tapper. This individual used to walk along a stationary train, striking the tires of the carriage or wagon wheels with a hammer; an apparently pointless activity. He was in fact practising one of the oldest of all forms of non-destructive testing. If a tire were cracked or otherwise defective, the sound it gave out on being struck differed from that given by a serviceable tire, and this difference could be detected by a practised ear. This at least was the theory, and smiths and ironfounders, among others, had attempted since time immemorial to assess the quality of a piece of metal by striking it and listening to the note produced. However, in spite of numerous attempts to refine and improve the technique, it was never a very sensitive one. The phrase 'as sound as a bell' has passed into the language, yet one of the world's most famous bells, Big Ben, is cracked in two places! It is heard every day by millions of people, few, if any, of whom can detect anything wrong from the sound of the bell.

It is from these crude beginnings, however, that the modern technique of ultrasonic testing has developed, though the old and the new methods have little in common apart from the use of sound. In order fully to understand the method, we shall have to look a little more closely at the nature of sound. If an object is struck, it will, of course, be set into vibration. This vibration is not necessarily confined to the object; it can be transmitted to an adjacent object, or passed on to the surrounding air, in which case it will spread, just as the waves set up by throwing a stone into a pond spread out from the centre. This vibration, or wave motion, travels at a speed which depends upon the medium—air, water, or anything else—in which it is occurring.

o

In air at sea level its velocity is 1,142 feet per second—about 730 miles per hour—in water, 4,900 feet per second, and in iron, 17,500 feet per second. Now any wave motion has, besides this *velocity of propagation*, three other important properties. These are frequency, that is, the number of vibrations, or cycles, which occur in a unit time, usually a second; wavelength, which is the distance from one wave crest, or peak, to the next; and amplitude, which is the height of the wave peaks. This is illustrated in Fig 26.

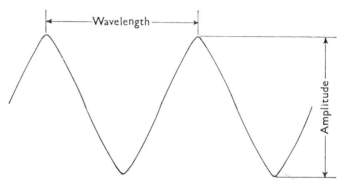

Fig 26   The definitions of wavelength and amplitude

The first three properties are related, wavelength being equal to the velocity divided by the frequency, or in mathematical terms, using the symbols conventionally assigned to these quantities:

$$\lambda = \frac{v}{f}$$

If these vibrations fall upon the ear they will, provided that their frequencies fall within a certain range, be perceived as sound. Frequency is experienced as *pitch*, thus, the note middle C on the piano has a frequency of 256 cycles per second (c/s), and the range of the piano is from about 25c/s for the lowest note to about 5,000c/s for the highest. The total range of human hearing is from 25c/s up to between 10,000 and 16,000c/s (usually expressed as 10 or 16 kilocycles per second, or kc/s). The upper limit varies between individuals, and depends to a large extent upon age. Some animals can hear much higher-pitched sounds than humans. Amplitude is perceived as loudness. There are, of course, other aspects of sound, such as quality, or timbre. These can all be explained in terms of the properties of the vibration, but for our present purpose it is not necessary to go into them. The sounds which we hear nearly always come to us through the air, and many musical instruments do in fact produce sound by setting a column of air into vibration directly. It is not essential that sound should travel through air, however; it does in fact travel rather better through solids and liquids. Were this not so, the problem of soundproofing would not be nearly so difficult as it actually is.

It is hoped that the reader now has an idea of what sound is. We must next see exactly what is meant by the term *ultrasonic*. Let us clear one source of confusion out of the way immediately; *ultra*sonic is not the same as *super*sonic, a term heard a good deal nowadays. The latter expression is descriptive of speed, and is commonly used to describe the speed of aircraft. A speed in excess of that of sound, which, as we have seen is about 730 miles per hour at sea level, is said to be supersonic. The term ultrasonic, on the other hand, relates to the frequency of vibrations, quite irrespective of their speed, and the ultrasonic frequencies are those above the normal range of the human ear, that is, higher than 16kc/s. The reason why vibrations of this order of frequency can be used for non-destructive testing depends upon the relationship between wavelength and frequency. The higher the frequency, the shorter the wavelength, and at ultrasonic frequencies, the wavelengths approach the size of the defects or flaws which have to be detected. We can easily prove this by considering the speed of sound in iron, already given as 17,500ft/s, and assuming a frequency of 100kc/s, which is quite a usual one for ultrasonic testing. Then, substituting these values into the formula on page 222, we have:

$$\lambda = \frac{17,500}{100,000}$$
$$= 0{\cdot}175 \text{ feet}$$

or about 2 inches.

If a sound wave of about this size, travelling through a solid medium, encounters some form of discontinuity, such as would be provided by a blowhole or crack, or some foreign matter, it will be partially reflected at the *interface*, that is, the junction between the defect and the mass of the specimen. Thus, if we know the amount of sound energy which has been put into the specimen, and the amount which would normally be absorbed in traversing it, assuming it to be perfectly homogeneous, we can calculate how much ought to be picked up by a suitable apparatus placed on the other side. If the amount actually received is less than this, we know that there has been reflection, in other words, there must be a defect in the path of the vibrations. By moving the receiving apparatus to a suitable position, we can detect the reflected energy, and so calculate the exact position and size of the defect.

So much for the theory. To put it into practice we obviously need a device to produce ultrasonic vibrations, and another capable of detecting them. Such devices are called *transducers*, because they convert one form of energy (usually electrical) into another, sound waves in the present case. Everyday examples of transducers working in the audible range are microphones and loudspeakers. Loudspeakers operate electromagnetically, and although ultrasonic vibrations can be produced in this way, it is not very convenient. Some microphones use a crystal to do the energy conversion, and this is the principle used in most ultrasonic transducers, both transmitters and receivers. It was pointed out on page 92 that if an alternating current is applied to certain kinds of crystals, they are set into vibration. With some crystals, this effect is very large, and the resulting vibrations are ideally suited to the purpose which we have in mind; by suitably shaping the crystal and using appropriate electronic circuits the frequency can be made anything desired, and the amount of energy applied, being electrical, is easily measured and controlled. Fortunately, the process is reversible; if a crystal picks up vibrations, it will generate an electric current of corresponding frequency. As a matter of fact, this is the principle of the crystal microphone just mentioned; it is also used in record players.

So we have our two transducers; now only one thing more is needed to make a workshop testing apparatus. The whole principle of the test is that energy is reflected at an interface; if the two transducers are simply placed on the specimen, it is obvious that two such interfaces are thus created, and a good deal of the energy will be lost.

Therefore it is necessary to have some means of *coupling* the transducers to reduce this loss as far as possible. This is usually done by interposing a layer of liquid: oil, water, and mercury are three which are commonly used, the exact choice depending on the circumstances of the test.

It will be obvious from this description that the results of an ultrasonic test, like those of neutron radiography, are in the form of a graph or a series of instrument readings (in practice they are often displayed on a *cathode-ray oscilloscope*), which have to be interpreted before they can have any meaning for the engineer. Thus a trained and skilled tester is needed, and there must be a proper programme of testing and a system of ensuring that it is properly applied. This indeed applies to all forms of non-destructive testing; the point is brought out here because, not so long ago, there was a serious railway accident caused by the breakage of an axle which was supposed to have been ultrasonically tested. The inquiry revealed that either it had not been so tested, or it had been put into service after having been found to be faulty. The reliance on tests is nowadays so absolute, as was demonstrated by this incident, that the importance of 'doing things by the book', and resisting the temptation to 'cut corners' cannot be over-emphasised. No matter how elaborate or ingenious test procedures may be, it is still necessary to provide for the human element.*

MAGNETIC CRACK DETECTION

One of the commonest, and also one of the most potentially dangerous, defects in metal parts, is surface cracking. This may be caused by fatigue, as mentioned in Chapter 7, but more often it is a factor contributing to it, and is itself due to some fault in manufacture, such as incorrect forging, or lack of proper precautions in grinding or heat treatment. The penetrant test is one way of detecting surface cracks, and is in general quite efficient, but it does require care in its application, and it takes time. Nowadays, unfortunately, time to do a proper

* This point was tragically reinforced in November 1967, when forty-nine lives were lost in a railway accident at Hither Green, just outside London. The cause was a broken rail, and the inquiry clearly showed that the fault had escaped detection because of a series of human failures.

job is the one thing that nobody seems prepared to allow, so a quicker method was sought for. To some extent, this demand is met by the method of magnetic crack detection. This can only be applied to components made of iron or steel, but this does, of course, cover a substantial proportion of machine parts and structural members made today.

It is common knowledge that the attractive power of a magnet (to put the matter in rather loose and unscientific terms) is concentrated at its poles. Furthermore, if a piece of iron or steel is placed within a magnetic field, for example by putting it inside a coil of wire in which an electric current is flowing, it will behave as a magnet. High-carbon steel continues to behave as a magnet even after the external field has been removed, but there are means of demagnetising it if necessary. Now, if an iron or steel component is magnetised in this way, a tiny pair of magnetic poles will be set up at any crack or flaw which exists on or near the surface. To make these visible it is only necessary to apply some magnetic particles, which will be attracted to the poles, and will thus reveal the outline of the crack quite clearly.

In practice, iron powder is often used, but magnetic iron oxide, which is the basis of magnetic recording tape, is even more effective, since it can be prepared in much finer form than metallic iron powder. The magnetic powder is often suspended in a suitable liquid. So, to detect cracks by this method, we simply place the part to be tested in a special rig which can apply a magnetic field, switch on the current, and either sprinkle the surface with the powder or coat it with the liquid suspension, which is usually referred to as 'ink'. A sharp rap will cause the superfluous powder to fall off, or it may be removed by more efficient means; the liquid will generally drain away of its own accord. In either case, any cracks which may be present will be immediately revealed by the powder or liquid which remains clinging round them. The test being complete, if the part is acceptable, the current is switched off, the part is demagnetised if necessary, and any remaining powder or liquid is wiped off. The whole test is the work of minutes.

Thus this test is very suitable for routine applications. It is not necessarily confined to the factory, for portable equipment has been developed which can be used, for instance, to detect cracks in welded rails without lifting them. This can be quite a consideration in modern railway practice, where up to a mile of track may be welded into a single entity.

ELECTRICAL RESISTANCE TESTING

Electrical resistance testing is to some extent related to the tests just described. If two pieces of metal are placed together, and arranged so as to form part of an electrical circuit, it is found that there is a greater resistance to the flow of current at the junction than in the body of the metal. In fact, if the join is not too perfect, and the current heavy, the metals will be heated almost to their melting points in the neighbourhood of the junction; this is the principle of resistance welding. Suppose, however, that instead of two pieces of metal with a junction between them, we have a single piece containing some kind of flaw, crack, or other discontinuity. Exactly the same effect will be obtained: the resistance of the piece will be greater than if it were homogeneous. It is theoretically possible for it to be less; for example, an unwanted piece of copper might somehow find its way into the middle of a bar of iron, but this is much less likely in practice.

If, then, a pair of electrodes is arranged so that they can be applied together to the surface of a part to be tested, the reading of a sensitive ammeter connected in series with them will be lower in the vicinity of a defect than it would be if only sound metal separates the electrodes. Thus, by systematically searching the entire surface, the defects may be localised. This method is not so sensitive as some of the others discussed, and it takes time, but it has the merit of being extremely simple to use—provided it is done conscientiously—and the apparatus is also simple and cheap. Thus the technique finds a certain amount of use.

PHOTOSTRESS ANALYSIS

To conclude this chapter, it is proposed to look briefly at a comparatively recent development, but one which is so elegant as to make it worth the slight effort involved in understanding it. First of all, it will be necessary to say a little more about the nature of light. We have already seen that this is a wave motion (the question of 'waves of what?' will continue to be dodged), that it has various wavelengths, which are perceived by the eye as colours, and that in general

227

its behaviour is subject to certain natural laws. Now, when we think about wave motions, we are apt to picture them either as waves on the surface of water, or, if we are considering a beam of light, rather in the manner in which waves are usually drawn in diagrams—such as Fig 26 on page 222. Neither image is particularly accurate. Even in water, the waves are not only upon the surface; they spread out in all directions, and this is true of waves in any other medium.

Waves of light and other electromagnetic radiations have another peculiarity: they do not normally vibrate all in the same plane, as is implied by the diagram mentioned; a more accurate representation would be something like Fig 27. This is not very easy to draw, so the other form is generally used for convenience. The representation of a beam of light in Fig 27 should be thought of as a 'bundle' of waves,

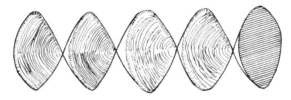

Fig 27   A sketch indicating the three-dimensional nature of the waves in a ray of light

vibrating in all possible planes at right-angles to the axis. Now it is possible to eliminate most of these, and obtain a wave motion of which Fig 26 would be an accurate picture, and light such as this, which is vibrating only in one plane, is said to be *polarised*. There are several ways of achieving this result, but the one most used nowadays is to pass the light through a piece of Polaroid. This is a synthetic plastic material which has a fine (ie, sub-microscopic) structure consisting of tiny crystals, long in proportion to their cross-section, and all aligned in the same direction. This structure acts as a kind of 'comb' or filter, and stops all light except that which has its axis of vibration in the same direction as the alignment of the crystals. If a second piece of Polaroid is now placed in front of the first, and the source of light is observed through the two of them, it will be found that the light-source cannot be seen unless the second piece is rotated to one of two quite definite positions 180 degrees apart. The reason for this is plain; these are the positions in which the crystal structure of the second piece is aligned with that of the first; in any other position the light will be cut off.

Polaroid will be familiar to many readers, as it is used in high-class sunglasses; its glare-preventing properties depend on the phenomenon just described, for the reflected light responsible for glare is itself quite strongly polarised. Now, what happens if we place a piece of some other transparent substance between the two Polaroids? Usually, nothing at all, but there are some substances (solutions of certain sugars are a good example) which cause the light to be cut off, and the second Polaroid has to be rotated by some amount before it is visible again. This proves that the substance introduced has had the effect of rotating the plane of polarisation. Certain other substances, however, give a much more spectacular effect. If they are introduced between 'crossed polars', that is, Polaroids set at 90 degrees, so that no light would normally come through, the most beautiful colour effects can be seen.

It is this latter phenomenon that we are interested in here. A full explanation of it would take up too much space, but the reader who has followed the discussion so far will realise that much more than a simple rotation of the plane of polarisation is involved. In fact, the effect is bound up with the internal structure of the substance, and this is what makes it useful from our point of view.

The concepts of stress and strain were explained in the last chapter. There only the simplest cases were dealt with; in testing materials it is desirable to keep things simple and avoid complicating matters unnecessarily, by eliminating, so far as possible, all factors other than the properties with which we are directly concerned. In real structures, however, the stresses and strains are by no means simple, and their distribution within the structure is a matter of vital importance in designing. In a component of a fairly simple shape, such as a beam or girder, the stresses and their distribution resulting from a given loading can be calculated; the mathematics involved is somewhat beyond the scope of this book, but is laborious rather than difficult. However, as the structure becomes more complicated, and the ways of loading it more numerous, the sheer weight of calculations involved soon becomes too much for most people, and to calculate all the stresses in a really complex structure is scarcely a practical proposition, even if a computer is available, and more direct methods are sought in practice.

The most usual method is to build a scale model and apply loads comparable with those to be expected in practice. The stresses and strains can then be measured by suitable instruments attached to various parts of the model, and for structures of appreciable size and complexity, this remains the best method. However, there is another way. It has been found that certain transparent materials, such as glass and some plastics, will exhibit colours between crossed polars if they are under stress, although in a normal, unstressed condition, they have no effect. Moreover, the colour patterns correspond exactly to the distribution of stresses in the article. One of the earliest practical uses of this discovery was in detecting 'locked-up' stresses in glassware. Glass objects are manufactured at very high temperatures, and they have to be cooled down to normal temperature very slowly indeed. This process is termed *annealing*. Any fault in the annealing process will result in the glass being under stress, in which condition it is liable to break in service conditions which would be withstood by properly-annealed glass. These stresses cannot be seen by simply looking at the glass, but they are immediately visible if it is examined between crossed polars.

It was not long before someone had the idea of making models of structures such as girders, bridge members, and so forth, from transparent materials, and examining them between crossed polars while they were subjected to various loads corresponding to those expected in the real thing. Glass is not ideal for this purpose, as it is quite difficult to fabricate, and it is brittle. The transparent plastics such as Perspex, however, are very suitable, and a model of the projected structure can quickly and easily be made and tested in the manner described. If the observed stress distribution at various loadings does not conform to what is required, it is a simple matter to make another model to a modified design and try again. Provided that certain simple rules are observed, scale models give an accurate reproduction of the stresses to be expected in the full-size article, but there is no reason apart from convenience and economy of material, why the models should not be made full size if desired.

This is a very effective technique, and one that has found a great deal of use; nevertheless it still leaves something to be desired. Although the stress distribution in a structure is largely dependent upon its shape, material can also influence it. Perspex is not steel, and cannot always be relied upon to behave in the same way. Moreover, the methods of construction, especially the means used to make the joints, will normally be quite different in the model and the real thing. Finally, and in some ways most important, this technique is of no use for testing an already existing object. There are ways of doing this, using 'strain gauges', but it is a laborious and time-consuming business, and the tendency is to apply it only in exceptional cases rather than as a routine measure.

These difficulties have now been overcome by the method of *Photostress Analysis*. This is a direct development of the technique just described; the only essential difference is that reflected, instead of transmitted, light is employed, but this one difference opens up a tremendous range of possibilities. Thus, if we illuminate some object with polarised light, and arrange to view it through a Polaroid screen, we can observe the stresses in it directly if only it can be made to behave like the transparent objects already described. The means of making ordinary opaque objects behave in this way is the heart of the method, and consists, quite simply, of a special coating which is applied to the test object. This coating is a fairly complex substance, and it is supplied ready prepared in several proprietary brands. The requirements are quite exacting. It is, of course, the coating itself which possesses the optical properties, but besides being compounded for these properties, it must be capable of bonding efficiently to a wide range of surfaces. Complete bonding is essential if the coating is faithfully to reproduce the stress conditions existing in the object to which it is applied. It must be easily and quickly applied, and is usually supplied in the form of a paint or lacquer; it must form a continuous and reasonably tough coating, and finally, it must be easily removed when it has served its purpose.

It is not altogether surprising that the manufacture of these coatings is a somewhat specialised business. However, results have amply justified the effort of developing and marketing them. The technique is a relatively new one, but it has already found hundreds of applications, and new ones are being reported almost every week. The photograph on page 204 shows, better than words can do, how the method works, and gives an idea of its applications.

9

Is it Real?

At the beginning of the last chapter, I promised to give an example of an early use of non-destructive testing, and this will now be done. As a matter of fact, this is probably the earliest example of non-destructive testing on record, and the story has often been told, but it may well be new to readers who have not previously taken much interest in science or technology. The name of Archimedes has already been mentioned as one of the few Greek philosophers who took much interest in experimental science. He actually lived in Syracuse, in Sicily, which at that time was a Greek dependency, from 287 to 212 BC. Many stories are told about him, most of them probably apocryphal, but the one with which we are now concerned is so good that one feels that if it isn't true, it ought to be.

As the principal 'wise man' of the place, Archimedes must often have been consulted about practical problems by the authorities, which in effect meant the king, Hiero, and possibly it was the need to please his royal patron, rather than any natural inclination, which led him to use experimental methods. One of the problems concerned the king's crown. This had been made to order, and for some reason the king suspected that the goldsmith had defrauded him by introducing some silver into the metal, which should have been pure gold. The technique of assaying (see page 234) was fairly well understood at the time, and no doubt the matter could have been settled in this way, but this would have meant destroying, or at best seriously damaging, the crown, and Archimedes was given the job of finding out whether the crown really was gold without damaging it. He pondered the problem for a long time, and was thinking about it one day in his bath. Incidentally, many people, including the present

writer, have found this a good place to think; nobody to my knowledge has ever suggested why this should be so.

Anyway, on this occasion, Archimedes noticed that as he stepped
into the bath (which would have been rather larger than those found
in present-day houses) the water rose, and pressed upwards on his
body. This gave him the clue he needed, and the story goes that he
was so excited that he leapt from the bath and ran naked through the
streets of Syracuse, shouting 'Eureka!' ('I have found it!'). Now what
in fact had he discovered to cause such excitement? He had realised
that when a body is immersed in water (or any other liquid, for that
matter), it will displace a bulk of liquid equal to its own bulk, but
that the weight of this bulk will not in general be equal to the weight
of the body. This is *Archimedes' principle,* usually stated thus:
'When a body is wholly or partly immersed in fluid, it suffers a loss
in weight equal to the weight of fluid which it displaces'. In other
words, if equal weights of gold and silver are weighed while immersed in water, both will appear lighter than if weighed in air, but
silver, being bulkier for a given weight, will suffer a greater loss than
gold. It should be noted, incidentally, that the water itself is not
weighed.

The application of this to the problem in hand is fairly clear.
Archimedes weighed the crown, and then took an equal weight of
gold known to be pure. He weighed this, and then the crown, when
each was immersed in water. In the second weighing, the crown
seemed to lose more weight than the pure gold, so it was in fact
composed partly of silver or something else. History does not record
whether he was suitably rewarded for the discovery, or what happened to the goldsmith, but, having regard to the penal code of the
time, one does not envy the latter!

The principle of Archimedes is an important one in physics, and it
has also been put to use for very much the same purpose as that
described. In fact, it is still very much used, as we shall see. This
testing of rare and precious objects was almost certainly the earliest,
and for long virtually the only, form of non-destructive testing. Gold-
smiths and jewellers have always used such methods, which should
not be confused with assaying. This is the most direct method of
finding out the composition of a metal sample; it involves an actual
analysis. One possible method of doing this has been described in

Chapter 4, but the methods of assay are many and varied. A frequent problem is to assess the metal content of an ore, to see whether it would be worth mining. Assaying is really a branch of analytical chemistry, but it also demands a considerable knowledge of metallurgy, and has nowadays become a specialised study in itself. One example of its use may be given.

Modern methods are very refined, and require only very small quantities of metal. This is most useful in assaying for hallmarking. A genuine gold or silver object will be found to be stamped somewhere with a series of cabalistic markings—letters, figures, and fanciful designs: a leopard's head, maybe, or a castle, a lion, a thistle, etc. These are the *hallmarks*; together they tell an expert the complete story of the piece. The first letters are the maker's mark, the others are put on by the assay office, to which all gold and silver articles made in or imported into Great Britain must by law be submitted before being offered for sale.

In the assay office, a tiny, and usually quite unnoticeable, scraping or clipping is taken from some inconspicuous part of the article, and this small amount of metal is enough for the assayer to tell whether the metal is of the required standard of purity. If it is, the hallmark is applied; if it is not, the article may be destroyed to prevent the public being deceived. This sanction, together with the very severe penalties for any misuse or copying of the marks, ensures that gold- and silverware offered for sale in Great Britain is guaranteed at least as far as the material is concerned.

Silver must be at least 92·5 per cent pure silver, which in happier times was the standard of the British coinage ('silver' coins nowadays contain no silver at all). There are several standards for gold, based upon the 'carat' scale. Pure gold, also termed 'fine' gold, would be 24 carat, but this would be far too soft for use, and will never be met with in an article of jewellery. Even 22 carat, 91·6 per cent gold, is still very soft, and 18 carat, 75 per cent gold, is the standard most often used for good-quality gold articles. The other standards are 14 carat, 58·5 per cent, and 9 carat, 37·5 per cent, this last being used for articles like rings and watch-guards which will have to stand up to a considerable amount of wear.

So much for that, but what of the jeweller faced with an article without a hallmark (possibly imported) or upon which he suspects the hallmark to have been forged (it has been done, in spite of the penalties). Well, first of all, in this matter of the examination of rare and precious things, more perhaps than in any other, trained and experienced judgement still plays a major part, a point which will later be stressed more than once. Thus, an expert can often detect at a glance something which to the layman appears to be a very passable imitation. Partly it is a matter of observation; most people 'go around with their eyes shut'. But take a nickel-plated and a chromium-plated article, and place them beside a piece of undoubted silver. They are all 'silvery' metals, but seen together the cold blue of the chromium and the cream tinge of the nickel are unmistakable against the real thing. (Note that 'real' in this context is a comparative term, and does not imply that chromium or nickel are inferior to silver; they may or may not be, depending upon the purpose in view.) Comparative weights, and the general 'feel' of the article will also tell much to the man accustomed to handling such things every day.

However, superficial examination will not solve every problem, and the jeweller will then most likely have recourse to a very old, but still valuable, aid—the touchstone. This is a piece of special stone, sometimes called Lydian stone. It is hard, fine-textured with a smooth surface, and is black or dark grey in colour. Some inconspicuous part of the article to be tested is drawn across the stone. It will leave a streak upon the stone; a small amount of the metal has actually been removed, but the quantity is so tiny that the test can be regarded as virtually non-destructive. For comparison, other streaks are made, using metal 'pencils' of known composition. Very often, the mere appearance of the streaks will disclose the required information to the expert eye, for the differences in metals and alloys seen thus are much greater than when they are examined as comparatively large masses.

However, to settle the matter definitely, further simple tests may be made. The various streaks may be touched with a drop of nitric acid, or *aqua regia*, which is a mixture of nitric and hydrochloric acids. The differing reactions of the various streaks to these substances will usually tell the expert what he wants to know. Thus, to take an extreme example, gold is unaffected by nitric acid, but brass would be dissolved completely, and the streak would vanish. It is unlikely that an experienced jeweller would have to proceed to the touchstone to distinguish gold from brass, however. Aqua regia dissolves most metals, including gold, but here the differing reactions give the information desired; colour changes may be evident, for instance. When the identification has been made, the streaks can all be cleaned off with aqua regia, and the stone is ready to be used for another test. Silver and silver-like metals can also be tested in the same way, only nitric acid being used in this case.

Nitric acid and aqua regia are the principal agents used for direct chemical tests of precious metals, although these are not so popular as the touchstone, as they may damage the article. They would be more likely to be used as a quick method of assessing the quality of something offered as 'old gold' or 'old silver' for remelting. Thus, a spot of nitric acid applied to an article containing some gold will show a greenish colour, with small bubbles, if it less than 9 carat. A pinkish-cream colour would mean that the article was of silver plated with gold. With 9 carat gold, a darkening effect would be seen, which would be less, and slower, if the object were of a higher carat, say 12 or 14. Aqua regia can be used on articles of over 9 carat; 18 carat gold and anything over would be untouched, whereas 14 or 15 carat gold would become paler.

The same reagents will reveal the difference between platinum and silver and similar metals. Nitric acid gives a greyish stain on platinum, but a cream one on silver; the latter will be darker and somewhat greenish if the metal is of lower quality. However, nitric acid dissolves silver, and would have to be used with care on an article not intended for breaking up. The standard test in such a case is to use a solution of silver nitrate. This has no effect on standard silver (often called 'sterling silver'), but will give a brownish stain on metal of lower quality; the greater the amount of base metal in the alloy, the darker will the stain be.

GEM TESTING

Precious metals are associated with precious stones in the minds of most people, so it is natural to consider next the testing of gemstones. Some, though not all, of the value placed upon gems is due to the fact that they are, in general, rare in nature, and to that extent, the value is artificial. This being so, and human nature being what it is, it is understandable that deception and self-deception have been practised in this field since time immemorial, and will probably continue to be. Incidentally, this seems to be a good place to explode one piece of 'folk mythology' which one frequently hears repeated: that it is not a good idea to take something to a jeweller for repairs in case he removes the diamonds (it is always diamonds) and substitutes less precious stones. In the first place, most of the very small diamonds forming part of a setting are not of particularly high value, and are certainly worth less than the time it would take to remove them and put in other stones. In the past, this trick may well have been played with larger stones of greater value, when the customer has been known to have more money than sense, but it is doubtful whether any jeweller would nowadays take such a risk, for a largish diamond is as identifiable as a famous portrait, thanks to the scientific methods of testing now available.

Gemmology is a science and a full-time study in its own right; it is a branch of crystallography, though it is also related to geology and mineralogy. To settle a knotty problem, it is often necessary to make use of the facilities of a laboratory, where a full range of testing equipment is available, with the trained staff to use it, but every jeweller needs a working knowledge of the subject, and can usually carry out at least the simpler tests. These tests may have one of two objects: to identify an unknown stone (and this may sometimes be a very difficult matter indeed), or, more usually, to determine whether something which is claimed to be a diamond, a pearl, a ruby, or whatever, is all that it pretends to be.

Most gemstones are hard; indeed, this constitutes one of their most valued properties. An early method of testing, therefore, was to check the hardness of the suspected stone, either with another stone of known genuineness, or with a file. This method is not much used nowadays, for it is very liable to damage the stone, which, even if not genuine, may very well be prized by its owner. One of the commonest everyday tests is to check the specific gravity of the stone. This can be done very accurately by a refinement of Archimedes' experiment described on page 233. The stone is weighed, using a very accurate balance, first in air, and then in water. The specific gravity can be calculated from the difference between the two weighings, and as the specific gravities of most of the gemstones likely to be met in practice

P

have been measured and tabulated in reference books, it is usually then possible to put a name to the stone, or at least to say whether or not it is what it is supposed to be.

This test, though accurate if carefully carried out, does take time, and the working jeweller often uses a less exact, but much quicker method. He will have a range of liquids of known densities. A stone put into one of these will float if its specific gravity is less than that of the liquid, and sink if it is greater. By a careful selection of the liquids, an identification good enough for most day-to-day purposes can be obtained.

The test just described can only be used for unset stones. There are, however, tests which can be used on stones which are still in their settings. Most of these depend on the optical properties of the stone; fortunately most gemstones are transparent or at least translucent. The phenomenon of refraction was explained in connection with the working of lenses in Chapter 6, where we saw how light is refracted as it passes from one medium, eg, air, into another, which was glass in the cases then considered. This refraction will be different for each pair of media, and can be expressed as a number; if one of the media is assumed to be a vacuum, a single *refractive index* can be assigned to every transparent substance. The refractive indices of nearly every possible gem mineral have been tabulated, so if we have a ready means of measuring it, this is an effective method of identification, for almost no two have the same refractive index.

The instrument used for the purpose is a *refractometer*. This works by comparing the refraction of the specimen with that of a specially dense piece of glass. The instrument is small enough to be held in the hand. In use, the gem to be tested, which in this case must be cut, with at least one flat facet, is placed with its flat face down on the glass table of the refractometer, a drop of special fluid being introduced between them. Upon looking through the eyepiece a shadow line is seen, and the position of this in relation to a numbered scale shows the refractive index. Indices between 1·3 and 1·85 can be measured; this range covers the majority, but not all, of the commoner gem minerals. Thus, diamond, with a refractive index of 2·42 is well off the scale. Incidentally, it is this comparatively enormous refractive index which accounts for the great brilliance, or 'fire' of the diamond.

When some stones, of which ruby, sapphire, zircon, and tour-maline are examples, are tested in this way, two shadows are seen in the refractometer. This is because these stones have *double refraction*, a single ray of light entering them being split into two rays, each of which is refracted to a different degree. These rays may actually have different colours, and this can be seen with another simple optical instrument, the *dichroscope*. This is fitted with an optical system which shows the two colours side by side.

Distinguishing true from false

The errors of identification, or downright deceptions, which may be met in dealing with gemstones are in the main of three kinds. First of all, one kind of gem may be mistaken for, or deliberately passed off as, another, more valuable one. Once his suspicions have been aroused, an expert will generally make a correct identification. He may have recourse to one or more of the tests already described, or possibly to some rather more complex ones, which we shall glance at in a moment, but very often he will reject the stone out of hand, a performance which appears to the layman as little short of witch-craft. In fact, of course, it is based upon experienced judgement and *careful observation*, which latter is the basis of all judgements of this kind.

Thus, a common substitute for diamond is zircon. Superficially, the resemblance is passable, but even without testing the hardness, which would be decisive, or using any instrument other than a hand lens, which is invariably brought into play when gems are being examined, an expert can frequently identify zircon by its double re-fraction. The facets may appear double when viewed through the back of the stone. Diamond, being singly refractive, would not give this result. Again, a simple test with the dichroscope suffices to dis-tinguish between real sapphire and blue spinel, a common substitute for it. Real sapphire is strongly dichroic, and shows two very distinct colours—deep blue and pale greenish blue—in the instrument, where-as blue spinel does not show this effect.

The second differentiation which may have to be made is between natural stones and synthetic ones. We saw in Chapter 5 that it is possible to make synthetic gems by high-temperature processes. Differentiation here is a much more difficult matter, for chemically, the synthetic and natural stones are identical, and possess identical physical properties to a large degree. Even the expert will hesitate to make a definite pronouncement in some cases without careful tests, and often the final verdict can only be given by a gemmological laboratory, which can use the more advanced methods shortly to be described.

The third possibility which may confront the jeweller is a manufactured imitation. It is in this area that the greatest variety will be found; ranging from the 'stones' used in cheap costume jewellery, and intended to deceive nobody, not even the layman, to the cunningly manufactured substitute, deliberately made to deceive even the expert, and consequently requiring all his skill to detect it. The commonest imitations are what the trade calls 'paste', which usually means some form of glass. Some very clever paste imitations have been made from time to time, but it is seldom that they will deceive even a working jeweller, provided only that he can handle them. One very simple, but impressive, test, is to touch the supposed gem to the lips, in company, if possible, with a real one. The basis of this—which the reader could try for himself if he is lucky enough to possess a fair-sized gem for comparison—is that glass is a poor conductor of heat whereas most gemstones are fairly good ones. Therefore glass feels warmer to the touch. The lips are used for the test since they are very sensitive to temperature differences. Any of the simple scientific tests described above would confirm the result.

Imitation gems have been made from transparent plastics, but their lightness and softness give them away at once. Another class of imitations are much more difficult to detect, for they are carefully made to pass as the real thing. These are called *doublets*. They may consist of two small gems cemented together to look like a larger, and therefore more valuable stone. These are rare nowadays, having been supplanted by synthetics. A doublet may also consist of a facing of real gem material cemented to glass. These may even pass an optical test, for instruments are nearly always arranged to test on the 'table' facet, that is, the large facet at the top of the stone as it is usually seen, and this is made of genuine material. The join is generally at the 'girdle' of the stone, where it is not easy to see. A careful specific gravity determination will often reveal these impostors for what they are, but again, in difficult cases, the laboratory may have to be called in.

Advanced methods

Let us now glance at some of the techniques which the laboratory will have in reserve for dealing with these difficult cases. First of all, an extension of the refractive index test to those stones outside the range of the refractometer is useful, and can be quite simply arranged —in fact, some jewellers use this method also. Just as it is possible to prepare a series of fluids having different densities, a similar set having varying refractive indices can be made up. If a transparent object, like a gemstone, is placed in a liquid having a refractive index fairly near to its own, it will be difficult to see, and if the indices are equal, the gem will 'disappear'. This test is usually done under the microscope, but a refinement of it, the *immersion contrast method,* perfected by Mr B. A. Anderson, enables the microscope to be dispensed with; moreover, a whole batch of stones may be tested at the same time, which is often very convenient. The stones are placed in a glass dish and immersed in a suitable fluid. The dish is then arranged so that its underside can be viewed in a mirror, and is illuminated from above. The differences between the stones show up in a remarkable manner, and can be still further enhanced and permanently recorded by making a direct photographic print. An example of such a photograph, made by the inventor of the process, is shown on page 204. It will be seen that the one real stone in the group of synthetic ones literally 'sticks out like a sore thumb'.

To the gemmologist, as to many other workers, the microscope is an indispensable tool. Even a hand lens will reveal that few gems are as perfect as they may appear to the naked eye, and the microscope will often show all kinds of flaws, gas bubbles, pockets of liquid, or solid bodies trapped within the stone. These are generically termed *inclusions*, and a careful study of them will tell the gemmologist much. In fact, the expert can often decide from them not only what the stone is, and whether or not it is genuine, but even where it came from. It must be understood, of course, that the inclusions we are now discussing are purely microscopic. Inclusions large enough to be seen with the naked eye would make the stone more or less worthless as a gem, since they would not only spoil the appearance, but would probably interfere with the cutting process. One exception is amber (which is actually a fossilised resin), in which the presence of gross inclusions, in the form of insects, was once held to prove the genuineness of the piece. Unfortunately, human ingenuity is only matched by human guile, and it is now possible to make artificial amber complete with insects!

Microscopic examination will often distinguish a synthetic from a natural stone. The usual process for making synthetic rubies and sapphires was described on page 152; it will be recalled that the stone is built up drop by drop. Very often, this leads to the presence in the stone of curved 'growth lines', not unlike the annual rings of a tree, but on a much smaller scale. These can, under suitable conditions, be seen in the microscope. Entrapped gas bubbles are another indication. However, the process of making synthetic gems is continually being refined, and detection by these means is becoming more difficult; indeed, it has been said that too great a perfection is nowadays a cause for suspicion. Of course, once the point is reached where the synthetic product is indistinguishable from the natural one, we may legitimately ask whether the latter is truly the more valuable, and whether the attempt to differentiate between the two is not a waste of time. These questions, however, are not for the scientist.

Possibly the most valuable ally of the gemmologist is the spectroscope. Most gemstones are transparent, so absorption spectra (see page 131) can be obtained by passing a ray of light through them and thence through a spectroscope. Practically the only difference between a ruby and a sapphire is the presence of a tiny amount of chromium in the one and of iron and titanium in the other. This difference can be demonstrated by the spectroscope, but except as a means of obtaining the knowledge in the first place, it is not necessary to do this, as the difference is immediately apparent to the naked eye. However, we may be confronted with two red stones; one may be a ruby, and the other perhaps a garnet, which is much less valuable. Though superficially alike, chemically they are quite different, and the spectroscope shows the difference quickly and unmistakably.

It is not, perhaps, surprising that stones formed in different parts of the world, though they may be very much alike, and would in general all be classified as 'rubies', 'emeralds', etc, do in fact differ very slightly in chemical composition. A spectroscopic test will often enable the gemmologist to say that a stone came from, say, Ceylon rather than Siberia, in a way that seems almost magical to the layman. However, where the spectroscope really comes into its own is in the rapid identification of the rarer and more obscure stones which are seldom met with in commerce. Spectra for most of these have been tabulated and published from time to time, and provided the data are to hand, the identification can be made in a matter of minutes, whereas by other methods it could well take hours, or even days.

It has already been mentioned, for example on page 215, that certain substances fluoresce when exposed to ultraviolet radiation. Some gemstones possess this property, and it often happens that stones which look very much alike when viewed in ordinary light, will fluoresce quite differently—perhaps one of them may not fluoresce at all—when irradiated with ultraviolet. The appearance of strongly-fluorescing stones is quite striking, and often very beautiful, but the value of the phenomenon to the gemmologist is that it provides him with another rapid, and sometimes decisive, test. Usually the stones are illuminated by a special lamp, the light of which is rich in ultra-violet, and which is provided with a filter of special glass which cuts off the visible light. Any fluorescence is then easily seen.

Another way is to illuminate the stones with any type of lamp which will give light rich in ultraviolet (usually a mercury-vapour lamp, of which the 'home sun' type of lamp is a good example) and examine them through a filter which allows light of the fluorescence colour to pass, but stops all other kinds. This filter, of course, provides a specific test for one kind of gem, and it would be necessary to have a whole range of them for general use, so the method is not used so much as the one first described. It does, however, form the basis of a test for distinguishing emeralds. These, if viewed through a special filter, even in ordinary light, look red, whereas other green stones with which they may be confused, remain green.

X-rays are sometimes used for examining gemstones. The idea in this case is not to see into the interior of the stone, in the manner described in Chapter 8, since most stones are transparent anyway. The technique used is 'X-ray diffraction', which depends upon the fact that the wavelength of X-radiation is smaller than the atoms of the stones. With a suitable set-up, crystalline materials irradiated with X-rays can be made to provide a 'diffraction pattern' on the photographic plate. This pattern is nothing at all like the actual crystal; it is representative of the arrangement of the atoms within it. Interpretation of the pattern requires some skill, but it can tell the expert a great deal; in fact, it could be described as the 'fingerprint' of the crystal.

X-ray examination is especially valuable in the case of pearls. The milky beauty of these has fascinated men (or, perhaps more accurately, women) since prehistoric times, and they often make a strong appeal to people who do not particularly care for other gems. Most people know that pearls are produced by certain molluscs, by far the majority of them coming from a certain species of oyster. Attempts have been made to imitate them, with varying success, for hundreds of years. As a matter of fact, however, imitation pearls present no real problem; they may pass muster with the layman, but an expert can identify them almost at sight. One quick test is to draw the suspect pearl across the teeth. Since the basis of imitation pearls is glass, they feel smooth when tested in this way, but a real pearl or a cultured one, feels rough.

Cultured pearls are produced by oysters, just like real ones, but the oyster is artificially induced to produce the pearl. Real pearls are rare, and therefore valuable, because first, very few oysters produce them, and a good deal of human labour, in the form of the difficult, and sometimes dangerous, work of pearl-diving, is needed to find them. Second, a fair proportion of those that are found are misshapen or otherwise faulty, and thus useless for jewellery—and there is no other use for pearls, as there is for some kinds of precious stones. Third, an oyster takes many years to build up a pearl of reasonable size. Incidentally, the reason why the oyster starts the process at all is something of a mystery. It is not invariably the case, as was once thought, that the pearl is built up around some foreign body. It may be, but it is not necessarily so. But once the process has started, it is continued year after year by the addition of successive layers of material, so the pearl gradually increases in size. This fact is very important in testing pearls, as we shall see in a moment.

Cultured pearls are the result of an attempt to overcome these difficulties. They are produced by taking young oysters and artificially implanting a bead of mother-of-pearl (this is a material taken from the inside of an oyster shell). The oysters are then returned to the water, and if all goes well, they start to add layers of pearl material to the bead. After a few years, when a thick enough shell has been built up around the original mother-of-pearl, they can be harvested. Not every oyster which is seeded in this way produces a pearl, but the proportion is much higher than is the case with 'wild' oysters. Moreover, if a pearl is produced, the shape of the bead ensures that its shape will be correct.

The testing of pearls, then, is almost entirely a matter of distinguishing between real and cultured ones. Sometimes the appearance in a strong light is enough for the really experienced eye, but some more scientific test is often desirable. X-rays are used, as mentioned above, and are especially valuable where the pearl has not been drilled. The difference in structure between real and cultured pearls shows up clearly on the diffraction photographs. This difference is also the basis of the *endoscope* test, which is in many ways more convenient than X-rays, and is certainly quicker.

This test is limited to pearls that have been drilled, but since by far the majority of pearls, real or cultured, are used for necklaces, the endoscope can nearly always be used. The general basis of its operation can be gathered from Fig 28, which shows the test diagrammatically, and on an enormously enlarged scale. The 'heart' of the endoscope is a needle, small enough to enter the drill-hole in the pearl, and which carries at its outer end two tiny mirrors, placed back to back and set at an angle. A narrow ray of light from a very powerful lamp is directed along the axis of the needle. The pearl to be tested is placed over the needle, while the observer watches through an eyepiece set in line with the needle and the drill-hole.

Fig 28　The working principle of the endoscope

If the pearl is a cultured one, the effect will be as shown on the left of the figure, where the heavy line represents the path of the ray of light. When the light encounters the first mirror, it will be reflected out of the pearl, and the observer will see nothing in the eyepiece, but a spot or a streak of light will appear on the outside of the pearl. If the pearl is a real one, however, the light, after reflection from the first mirror, will pass around the concentric layers in the manner shown, until it encounters the second mirror, when it will be reflected along the original line and into the eyepiece. The observer sees this as a distinct and unmistakable 'flash'.

This test is virtually infallible, and very quick; some hundreds of pearls can be tested in a working day. Unfortunately, the complexity of the instrument, and even more, the difficulty of making the needles, mean that it is a laboratory instrument rather than an everyday test.

OTHER VALUABLE OBJECTS

We have already seen that considerable ingenuity has been devoted to making replicas of, or substitutes for, things which are found only rarely in nature, and it has been implied that much of this effort is due, in part at least, to the rather artificial values placed upon these objects. It is not surprising, then, that man-made objects which are for some reason or other considered to be valuable, should also have attracted the attention of counterfeiters. In general, the objects of this class which are most often imitated are those with but little intrinsic value, and whose value stems, broadly speaking, from originality or association. Neither rarity nor workmanship as such have very much to do with the matter (though, of course, many such objects are both rare and well made). This can be shown by considering that a catalogue of any given nineteenth-century firm is probably rarer than a first folio of Shakespeare (it is so, incredible though it sounds), further, the latter is a distinctly poor example of printing, and much better copies can be obtained for a few shillings. However, the logic, or otherwise, of the valuation does not really concern us here; given that something, if real, is valuable, we want to know how its genuineness can be established.

A surprising amount can be found out about anything simply by a thorough and informed examination. There is no doubt that most people are not nearly as observant as they might be, and one difference between the expert and the layman is that the former is trained to see, not merely to look. This by itself, however, is not enough; practice and experience are also necessary. Often this knowledge, though it may be possessed, is half-unconscious; thus, in the days when craftsmanship was much commoner than it is today, a workman could usually say whether or not he had made some object, and he could often identify the actual maker, if it was not himself, though perhaps without being able to say exactly why he was so certain. Possibly this is the mark of the real expert; he can usually put the reasons for his judgements into words. Without this ability, any such judgement must remain subjective and valueless; whether it is right or wrong is irrelevant.

The principle behind this examination is to be found in the performance of the craftsman mentioned above. Since (so far as anyone knows) no two people are exactly alike in physical and mental make-up, education, upbringing, and general experience of life, it follows, more or less inevitably, that nothing which two different people make or do will be exactly the same. This is most commonly seen in handwriting, and indeed, to some extent our entire legal and commercial system is based upon the fact. Thus, a bank is held to be responsible for money paid out on a forged check; the law considers that the bank has an absolute duty to know its customer's signature, and it is not accepted that even the cleverest forgery can be exactly the same.

This individuality is in fact most strikingly demonstrated by the signature. Whether you write your signature on paper with a pen, pencil, or brush, or trace it in sand with a finger-tip, it will still be the same, and like nobody else's. It would still be the same if you had the skill (some people have) to cut it in steel with a hammer and chisel! Thus, the expert seeking to establish whether a painting, sculpture, piece of furniture, or any other artifact is genuine, looks primarily for these marks of individuality which are exceedingly difficult to reproduce. The artist's manner of using his brush, the carver's preference for a favourite chisel, the cabinet-maker's grip of his plane, are all individual, and all leave their mark on the finished work to tell a story to those who have the knowledge to read it, just as does the writer's choice and arrangement of words. Incidentally, the latter can also be analysed nowadays, though this is a job for a computer.

To observe these marks of individuality is one thing; to reproduce them is quite another. It has been done, however, though seldom well enough to deceive *all* the experts. One instance where this did almost happen was in the van Meergeren case. Van Meergeren was a Dutchman who after the Second World War was accused of selling valuable works of art to the German occupation forces. His defence, that the paintings he had sold were his own work, was ridiculed at first, for art experts testified that the paintings were genuine 'old masters'. The matter was settled beyond doubt, however, when van Meergeren actually painted a 'Rembrandt' under strict supervision, and the experts retired discomfited. This was a case in which a few simple scientific tests would have settled the matter out of hand, and saved quite a few reputations. Most art experts have profited from the example, and will nowadays invoke the assistance of the laboratory before finally deciding in a difficult case.

Most laboratory tests of paintings, furniture, and other *objets d'art* depend upon the ability to analyse tiny quantities of substance, and are based upon the fact that the materials available to the modern forger are, in general, not the same as those used by the original worker perhaps two or three hundred years ago. This affects workers who have no intention of producing imitations; for example, the modern blacksmith making ornamental ironwork cannot nowadays obtain the wrought iron which is the traditional material, and has to use mild steel. Much the same considerations apply to the compounding of pigments and paints, and since no forger is likely to have unrestricted access to the original, together with the advanced laboratory facilities needed to make an analysis of the pigments, he starts under a severe disadvantage.

A rapid non-destructive test of a painting can quite often be made by photographing it on a plate sensitive to infrared radiation (see page 133). Such a photograph usually looks quite different from one taken by ordinary light, but the interesting and useful thing is that the 'straight' linseed oil varnishes used in old paintings photograph quite differently from more modern ones.

Where the facilities are available, it is a comparatively simple matter to take small samples from the suspected picture and from an undoubtedly genuine one, and compare them by using the techniques of spectroscopy and chromatography described in Chapter 4. The nature of the original materials is not the only tell-tale factor; all paints undergo chemical changes when they dry, such changes being in fact the basis of the drying process. It is obvious that these changes may be expected to have proceeded farther in a really old painting than in a comparatively modern one. It has indeed happened that the chemical changes in the paints have been such as to endanger the very existence of the painting—some of the 'old masters' seem to have been rather careless in their colour mixing—and remedial action has had to be taken. The fascinating laboratory techniques involved in the business of restoration are, however, outside the scope of this book.

Similar methods can be applied to metal objects, and the tiniest scraping, even if it is corroded, provides sufficient material for a paper chomatogram. Coins and other metal objects from Egypt and other early civilisations have been tested in this way; in these cases there was no doubt of their genuineness, but the results of the analyses permit of deductions regarding the methods and materials used by the early metalworkers—information which it would be hard to obtain in any other way. Further, the data being on record, are useful for the detection of any forgeries that may be attempted. Forgeries are not unknown, particularly of Egyptian relics. Leonard Cotrell has recorded that many of the 'antiquities' found in and around the Necropolis at Thebes were in fact manufactured by the descendants of the original craftsmen who still live in the area. The techniques must have been handed down over something like 2,500 years since the last Pharoah was buried there—an astonishing survival.

Pottery can also be tested in this way, since the constitution of the raw materials in various parts of the world differs quite considerably. Stone sculpture would present greater difficulties, but fortunately it does not appear to have attracted the attention of the 'faker' to any great extent. The reason is fairly obvious; a good deal of work is involved, and it is not so easy to 'pass off' sculpture. A painting may plausibly be represented as having been 'lost' for many years, and many such instances, involving genuine works, are in fact on record. But it is much less likely that any of the work of the great sculptors could have remained in obscurity, and there is no point in faking antiquities—anyone capable of producing a sister to the Venus de Milo would be only too proud to present it as his own work.

One can, however, never be quite sure, as was proved, in a closely related field, by the exposure of the Piltdown skull some years ago. This object, supposed to be the skull of a primitive form of man, was discovered in a gravel pit in Sussex in 1912. About forty years later, laboratory tests at the Natural History Museum, London, showed it to be a 'fake', manufactured with astonishing skill and ingenuity, for reasons that can only be guessed. It is, however, fair to say that this particular 'relic' had always been regarded with a good deal of suspicion, and it had never been placed on public exhibition for this reason. The scientific tests merely confirmed the earlier doubts.

There is not usually much point in applying chemical tests of this kind to old furniture, for it will normally have been repolished more than once in its lifetime. Provided that suitable specimens can be taken, the age of wood and indeed of most organic materials can often be established by microscopic examination. However, counterfeiters are sometimes clever enough to use old wood in their forgeries, so the expert's judgement based on considerations of style and workmanship is still called for. Provided that an estimate of age is all that is needed, and that a sample of the material can be sacrificed for testing purposes, a fairly accurate dating can be made by the radiocarbon method. Limitations of space, however, must very soon bring this book to an end, and we cannot go into details of this technique.

Perhaps the one class of objects which more than any other has received the unwelcome attentions of forgers over the years is documents, with which of course we must include checks and banknotes. The problem of forgeries of these latter is tackled from two angles; first the checks or notes are so made as to be very difficult to copy, and also it is arranged that, so far as possible, forgeries shall be relatively easy to detect. When a document is alleged to be old, but in fact is not, chemical tests on the lines described above will often reveal the fact. More than one forger has been caught because he overlooked, or did not know, that aniline dyes, which are used in most modern inks, were not invented until the latter part of the nineteenth century, and prior to that time, inks were based upon tannin (tannic acid), mostly obtained from oak-galls, and an iron salt. Attempts to 'age' paper by staining it are equally easy to detect in the laboratory.

However, perhaps the most valuable weapon of the scientist who specialises in this kind of work is the various invisible radiations which have been described earlier. Thus a document photographed by ultraviolet radiation will often reveal not only that it has been altered, but will show the original writing. Infrared radiation can be similarly used, while X-rays also have their uses. The latter have shown how one painting has sometimes been made over the top of another. Often in these cases, as also with old documents that have been erased and used again, there is no question of forgery, but the information so revealed may be of great historical interest. To go into this in detail, however, would practically involve starting another book, whereas it is now time to bring the present one to a close. However, it may be hoped that the reader whose interests incline him towards this specialised, but fascinating branch of the subject, may be moved to pursue his studies in greater depth elsewhere.

Reading List

The purpose of this list is to provide starting points for the study in greater depth of the subjects discussed in this book; it is not intended to be a comprehensive or up-to-date bibliography. So far as is possible, a good popular treatment of the subject in question, where such is known to the author, is given first, followed by one or more texts at a rather more serious level. As it is desirable, at least in the initial stages, that the student should possess his own books for reference, every effort has been made to include books which are cheap and easily obtained—many of them can be had in paperback editions. However, a few excellent works which are currently out of print are included; most of them can be borrowed from the public libraries.

Chapter 1 and general

CROMBIE, A. C. *Augustine to Galileo*, 2 vols, Heinemann, 1959, also in paperback, 1961; Harvard University Press, 1961. A paperback edition is published in the USA by Doubleday-Anchor under the title *Medieval and Early Modern Science*
The historical approach to the study of science is often recommended to the non-scientist. This book gives an excellent account of the development of the scientific method from this point of view.

DERRY, T. K., and WILLIAMS, T. I. *A Short History of Technology*, Oxford, 1960
Very useful for anyone wishing to gain a synoptic view of the development of technology. Does not assume too much technical background.

ISAACS, A. *Introducing Science*, Penguin, 1963

A sound, though necessarily limited, general introduction. Suitable for those who have not previously studied scientific subjects.

Chapter 2

ANDERSON, R. W.   *Romping through Mathematics*, Faber, paperback edition, 1960
An admitted lightweight, but one which contains a great deal of information, and an excellent starting point for anyone with a dread of the subject.

HOGBEN, L.   *Mathematics for the Million*, 3rd edition revised, Allen & Unwin, and Norton, 1951, and numerous reprints
This was probably the first book about mathematics written for the complete layman. It contains a tremendous amount of material, but it cannot be pretended that the author's determined and conscious concentration upon a 'historical' rather than a 'logical' approach makes it an easy book. Also, his tendency to drag in his political views will undoubtedly annoy many readers, and tends to detract from the book's value. Nevertheless, it cannot really be excluded from this list.

SAWYER, W. W.   *Mathematician's Delight*, Penguin, 1943, and numerous reprints
Another 'layman's introduction', much less comprehensive, but far more readable and more logically arranged than Hogben (above). Recommended.

MARTIN, G. C., and SMALLEY, A.   *Basic Mathematics*, English Universities Press, 1963
A 'programmed learning' course, designed for private study. It deals only with algebra, but anyone who follows it to the end should have a reasonable knowledge of, and a fair proficiency in, the subject.

RAWLINGS, G. P.   *Trigonometry Made Plain*, Percival Marshall, 1948
A very sound work, with an intensely practical approach.

RAWLINGS, G. P.   *The Calculus—Arithmetic of the Age*, Percival Marshall, 1951; republished by Model & Allied Publications, 1964
One of the best simple introductions to this branch of mathematics.

Reading List

ABBOTT, P., KERRIDGE, C. E., MARSHALL, H., and MAHON, G. E. *National Certificate Mathematics*, 3 vols, English Universities Press, 1938 and many reprints

A standard textbook for those wishing to embark on serious studies. Written primarily for technical students, the approach is practical rather than theoretical.

MORONEY, M. J. *Facts from Figures*, 2nd edition, Penguin, 1953 and reprints

This is described as 'a layman's introduction to statistics', which indeed it is, but in soundness of approach, the amount of material contained, and clarity of exposition, it is far superior to many much weightier and more expensive volumes.

Chapter 3

IRWIN, K. G. *Man Learns to Measure*, Dobson, 1963

An original treatment, primarily aimed at the younger reader, but will appeal to many adults.

DE CARLE, D. *Teach Yourself Horology*, English Universities Press, 1965; Dover Publications Inc

An introduction to time measurement. The author is an acknowledged authority, responsible for many more advanced works.

FEATHER, N. *Mass, Length and Time*, Edinburgh University Press, 1959; Penguin, 1961

A serious work, which deals with much more than measuring methods. Requires some knowledge of mathematics.

Chapter 4

HUTTON, K. *Chemistry—the Conquest of Materials*, Penguin, 1957

An excellent introduction to chemistry for the general reader.

KNIGHT, J. *Teach Yourself Chemistry*, revised by G. B. McAlpine, 3rd edition, English Universities Press, 1966

A sound introduction to experimental chemistry for the private student. The nomenclature is somewhat old-fashioned, but its emphasis on experiment is undoubtedly right, and it is written with the needs of the home worker in mind.

READ, J. *A Direct Entry to Organic Chemistry*, Methuen, 1948 and reprints

One of the finest pieces of popular science writing of the post-war years. The author attempts a nearly impossible task, and succeeds brilliantly. No previous knowledge is assumed; the book is extremely readable, yet scientific accuracy is nowhere sacrificed, and a tremendous amount of information is included. Highly recommended.

GIBBS, F. W. *Organic Chemistry Today*, Penguin, 1961
While not in the same class as Read (above) for exposition, this book mentions an enormous number of substances, and is a useful reference work. Contains a good bibliography.

Chapter 5

There is no single work dealing with the subject of attaining and measuring high temperatures. The matter is dealt with in more or less detail in works on chemistry, metallurgy, physics, etc, according to need. For low temperatures, refer to:

MENDELSSOHN, K. *The Quest for Absolute Zero*, Weidenfeld & Nicolson, 1966; McGraw-Hill
Here the story is told by one of the foremost present-day workers in this field, who is able to relate much of it from personal experience. Some general acquaintance with scientific and technical matters is assumed; it should be within the scope of anyone who has read the present book.

Chapter 6

MILLER, R. F. E. *The Amateur's Microscope*, Percival Marshall, 1951
A good general introduction. Contains a chapter on constructing one's own equipment, for which, however, a fair knowledge of, and skill in, metalworking is required.

WOOD, J. G. *Common Objects of the Microscope*, rewritten by W. J. Ferrier, Routledge, 1938 and reprints
This is one of the classics of popular science. It is now more than a hundred years since it was first published, and its longevity is sufficient testimonial to its value. Although the text has been revised and rewritten more than once, to keep it up to date, the beautiful coloured drawings—examples of an almost lost art—which are the central feature have been retained in the modern editions.

Martin, L. C., and Johnson, B. K. *Practical Microscopy*, 2nd edition, Blackie, 1949 and reprints
The title is sufficiently descriptive; this is a handbook dealing with the technique of using the microscope.

Drew, J. *Man, Microbe, and Malady*, Penguin, 1940 and reprints
A simple introduction to the science of bacteriology. Has some good photomicrographs.

Chapter 7

Practically all the available works covering the subject of this chapter are addressed to those who already have a fair amount of technical knowledge. A brief sketch is presented in:

Street, A., and Alexander, W. *Metals in the Service of Man*, 4th edition revised, Penguin, 1968
This book, however, is chiefly useful as a supplement to the present one: it provides a good deal of information about metals and metal-working. Introductions to testing can be found in some general works on mechanical engineering, and such standard references as:
Kempe's Engineer's Year Book, 2 vols, Morgan Bros, annually.
A much more specialised and comprehensive survey is:

Fenner, A. J. *Mechanical Testing of Materials*, Newnes, 1965
This assumes a knowledge of engineering as a matter of course, but may be found useful for reference.

Chapter 8

Fordham, P. *Non-destructive Testing Techniques*, Business Publications, 1968
This book is claimed to be addressed primarily to business executives and accountants who may not possess specialised technical knowledge, and the supposed needs of this readership may possibly account for the rather heavy reliance upon manufacturers' published material. It will certainly be intelligible to a reader of the present book, who, however, should then progress as soon as may be to the next work mentioned.

Lamble, J. H. *Principles and Practice of Non-destructive Testing*, Heywood, 1962

Reading List

This gives a fairly comprehensive and not too difficult account of all the various methods which were in use up to the time of its publication.

Chapter 9

SELWYN, A. *The Retail Jeweller's Handbook*, 7th edition revised by
    J. J. Adler and G. F. Andrews, Heywood, 1964

This is hardly a book which the general reader will wish to buy, for, as the title suggests, it contains a great deal which is of interest only to those 'in the trade'. However, it gives clear and not too technical descriptions of the methods of testing used by working jewellers.

ANDERSON, B. A. *Gem Testing*, 6th edition, Heywood, 1965;
    Emerson, 1965

This is not a textbook of gemmology, though it gives references for the benefit of those wishing to study the subject further. Its field is exactly defined by its title, and it is probably the only book dealing exclusively with this field. Intended largely for the working jeweller, the approach is practical, but with a very sound scientific basis. It is perhaps worth remarking that many amateurs have been attracted to the study of gemmology, which is not as expensive as might appear, and which provides a good grounding in several scientific disciplines.

The books listed above were all published originally in the United Kingdom, though many of them are also obtainable in the United States and elsewhere, sometimes under different imprints, as indicated. However, readers in the USA and some other countries may find more convenient or more congenial starting points for further study among the following list of books. These books are listed in alphabetical order of authors' names.

ADLER, IRVING *Hot and Cold*, John Day, 1959
    A New Look at Arithmetic, John Day, 1964. Also in paperback;
    Signet, New American Library
    A New Look at Geometry, John Day, 1966: paperback, Signet
    The New Mathematics, John Day, 1958: paperback, Signet
    Probability and Statistics for Everyone, John Day, 1963: paperback, Signet

Reading List

ALLEN, R. J. *Cryogenics*, Lippincott, 1964

ASIMOV, ISAAC *An Easy Introduction to the Slide Rule*, Houghton Mifflin, 1965: paperback, Fawcett World Library
The New Intelligent Man's Guide to Science, Basic Books, 1965
The Intelligent Man's Guide to the Physical Sciences, Washington Square Press, 1968
Of Time & Space & Other Things, Doubleday, 1965
A Short History of Chemistry, Doubleday-Anchor, 1965
The World of Carbon, Abelard-Schuman, 1958; paperback, Collier-Macmillan
The World of Nitrogen, Abelard-Schuman, 1958; paperback, Collier-Macmillan

BITTER, FRANCIS *Mathematical Aspects of Physics*, Doubleday-Anchor, 1963

BOYD, W. T. *The World of Cryogenics*, Putnam (NY), 1968

COSSLETT, V. E. *Modern Microscopy; or, Seeing the Very Small*, Cornell University Press, 1966

CROWLEY, T. H. *Understanding Computers*, McGraw-Hill, 1967

FINK, DONALD G. *Computers and the Human Mind*, Doubleday-Anchor, 1966

FISHLOCK, DAVID *Taking the Temperature*, Coward McCann, 1968

HALL, J. A. *The Measurement of Temperature*, Barnes & Noble, 1966

LEY, WILLY *The Discovery of the Elements*, Delacorte, 1968

LIEBER, LILLIAN R. *Take a Number: Mathematics for the Two Billion*, Ronald, 1946

REID, R. W. *The Spectroscope*, Signet, New American Library

Articles on a wide range of scientific and technical subjects are also regularly published in certain scientific journals which specialise in an 'interdisciplinary' approach. *New Scientist, Science Journal,* and *Scientific American* have international circulations, and many of their articles, while being invariably sound in content and serious in presentation, are very readable and well within the comprehension of an interested non-scientist.

Appendix

A Note on the SI Metric System

Since Chapter 3 of this book was written, there has been a gradually increasing public awareness of the implications of the official decision to change over to the metric system in Great Britain, and 'metrication', as it has been termed, has become a topic of some current interest. As it happens, the decision to make the change more or less coincided with the publication of the latest proposals of two international bodies, the Conférence Générale des Poids et Mesures (General Conference on Weights and Measures) and the International Organization for Standardization (ISO), concerning a system of metric units which had first been introduced in 1960. This is the Système International d'Unités (International System of Units), for which the international abbreviation is SI, and the proposals having met with general approval, several countries which already use the metric system began to prepare legislation to make the SI the only legal system. It was obviously logical that the United Kingdom, making the change to the metric system at this time, should adopt the SI units from the outset, and this is in fact being done. A brief outline of the system may therefore be of interest. Only the barest sketch can be given here, and readers are strongly advised to obtain a copy of publication PD 5686 of the British Standards Institution, *The Use of SI Units*, which deals with the matter in greater detail.

The SI is based upon the system described in Chapter 3 as the MKS system. It is not quite a 'pure' MKS system, in which all units would theoretically be defined only in terms of the metre, kilogram, and second, but it is a coherent system, which means that if any two unit quantities in the system are multiplied or divided, the product or quotient will be the unit of the resultant quantity. Thus, unit length (the metre) multiplied by itself, gives unit area (the square metre), unit length divided by unit time (the second) gives the unit of velocity (metres per second) and so on. The system consists of six *base units*, two *supplementary units*, and a number of *derived units*.

The base units are arbitrarily defined, as explained in Chapter 3: they are, with their official abbreviations, as follows:

metre (m), unit of length
kilogramme* (kg), unit of mass
second (s), unit of time
ampere (A), unit of electric current
kelvin (K), unit of thermodynamic temperature
candela (cd), unit of luminous intensity

Of these, we are already familiar with the metre, kilogramme, second, and ampere; the kelvin is simply the 'degree Kelvin', identical with a Celsius degree, and introduced in Chapter 5, but it is no longer necessary to say or write the word 'degrees' when working with the SI units—'kelvin' suffices. The candela was originally defined as the luminous output of a certain area of platinum solidifying from the molten state.

The supplementary units are the radian (rad), the unit of plane angle; and the steradian (sterad), the unit of solid angle. The radian has long been in use as a unit of angle for scientific and technical purposes; it is defined as the angle subtended at the centre of a circle by an arc equal in length to the radius, therefore 2π radians = 360 degrees. One steradian is the solid angle subtended at the centre of a sphere by an area on the surface numerically equal to the square of the radius. Quite a strong visual imagination is needed to picture this last, but it has important practical uses.

The derived units are expressed in terms of the base units; thus the unit for velocity is metres per second (m/s) as mentioned above. Some of the commoner ones have been given special names and symbols, a few of which were mentioned in Chapter 3. Thus the unit

* This is the 'official' spelling, but 'kilogram' has been adopted throughout the text of this book as being more in line with English and American orthography.

of force is called the newton (N); of energy, the joule (J); and of power the watt (W). The adoption of a unit of force, incidentally, does away with the necessity of using weight units to express force, which has always been open to objection, primarily on theoretical, but also on practical grounds. Thus pressure, for instance, in SI units, is expressed as newtons per square metre (N/m^2). There is no space here to set out all the derived units in detail, but one which might be mentioned is the unit of frequency, the hertz (Hz) equal to $1/s$. Thus, the standard alternating current supply in Great Britain has a frequency of 50Hz.

It is obvious that given a coherent system of units, work will be saved and many errors avoided, by always carrying out calculations in terms of the unit quantities, and it is one of the principles of the system that this shall be done. However, for practical purposes, it is often necessary to retain a system of subdivisions of the units, and for this purpose the decimal prefixes of the 'classical' metric system, together with a few more modern ones, are employed, but it is recommended that as far as possible only those prefixes representing ten raised to a power which is a multiple of three should be used. Thus the 'preferred' prefixes are:

Factor	Prefix	Symbol
10^{12}	tera	T
10^9	giga	G
10^6	mega	M
10^3	kilo	k
10^{-3}	milli	m
10^{-6}	micro	μ
10^{-9}	nano	n
10^{-12}	pico	p
10^{-15}	femto	f
10^{-18}	atto	a

The symbol for the prefix is combined with the symbol for the unit concerned, and the combination is then regarded as a single unit symbol. Examples with which we are already familiar are mm and km. The units of measurement hitherto called the 'micron' and 'millimicron' (see page 74) now become micrometre (μm) and nanometre (nm) respectively.

Appendix

Compound units are ruled out, which has led to some difficulty over the units of mass. For example, 10^{-3}kg should logically be written mkg, but this is confusing, and so the names and symbols gram (g), milligram (mg), etc, have had to be retained. No doubt the best course would have been to have renamed the kilogram, but it was so familiar to so many people that this was hardly practical.

It is recognised that these rules, though they are completely rational and well adapted for scientific and many technical purposes, cannot be applied in their entirety for everyday purposes. For example, division of time by weeks, days, hours, and minutes will continue, even though scientists may speak in terms of kiloseconds, etc. Likewise, the centimetre, though not a 'preferred' unit, is such a familiar and useful one in continental countries that it is unlikely to be discarded, though the British, coming fresh to the system, may be able to do without it. To the average man, thinking mainly in terms of journey times, kilometres per hour (km/h) is also likely to be more meaningful than metres per second.

The comparative inaccuracy of volumetric measurements was pointed out in Chapter 3, and the SI originally made no provision for volumetric units except as derivatives of the metre and its subdivisions. However, for everyday purposes the litre is now admitted, being regarded as a synonym for the cubic decimetre (dm^3), though the use of the name litre for precision measurements is discouraged. The litre and millilitre (ml, equal to $1cm^3$) should be adequate when the United Kingdom 'goes metric', and there would appear to be no good reason for taking over the decilitre so popular in continental cookery books.

Temperatures for everyday purposes will also continue to be expressed in °C, but since the Celsius degree and the kelvin are identical, there is no essential conflict here.

Index

Page numbers in italic type refer to illustrations

263

Index

Microscope, electron, *185*; results obtained with, 182, *186*; scanning, 182; working principles, 180, *181*

Microscope, optical, 210, 241; binocular, 174; compound, 169–74, *172*; condenser, 171, *172*; eyepiece, 171, *172*; general-purpose, 171, *167*; high-powered, 171, 173; metallurgical, 173, 191, *192*; objective, 171, *172*; petrological, 173; portable, 174; resolution of, 171, 179; simple, 169

Microscopy: of bacteria, 178–9, *168*; of dust, 177, *168*; of fibres, 178, *168*; of gemstones, 241; of metal specimens, 190, 193–4, *193*

Mohs' scale, 205

Molecular weight, 122

Molecule, 108

Multiplication law, 46

Newton's rings, 211, 212

Nitric acid, 235, 236

Non-destructive testing (NDT), 208–31

Objets d'art, testing of, 248

Observation, 17, 235, 239, 246

Ohm's Law, 97, 98

Oil-and-whitewash test, 214

Operators, mathematical, 34

Ordinates, 39

Oxygen, 145, 146, 158, 159

Oxy-acetylene blowpipe, 146

Oxy-hydrogen blowpipe, 146, 152

Paintings, laboratory tests of, 248

Palmer, Jean, 85

Palynology, 177, 178

Parameters, 33, 95

Pearls: cultured, 244–5; testing, 245–6; X-ray examination of, 244

Peltier effect, 163

Pendulum, 89, 90; free, 91

Penetrant testing, 214, 215

Perception, 17

Periodic table, 109; discovery of, 110; modern form, 111, *112*

Permutations, 47

pH meter, 118

pH scale, 118

Philosophers' Stone, 103, 104

Photomicrography, 174–6, *168*

Photostress analysis, 227, 231, *204*

Physical change, 105

Piltdown skull, 249, 250

Pinched plasma discharge, 147

Pipettes, 78, *79*

Planimeter, 54, *54*

Platinum, identification, 236

Pneumatic gauging: of bores, 86–7; of crankshaft, 87, *66*

Polarization of light, 228–9

Pollen, 177, 178, *168*

Polymers, 187

Power, 94

Prediction in science, 21, 22

Pressure tests, 213

Probability, 45–7

Prony brake, 95, *96*

Punched card principle, 57

Pyrometer: expansion, 142, *143*; optical, 144; thermocouple, 144

Pythagoras' Theorem, 31; proofs of, 31

Quality control, 26

Quantitative work, 27

Radicals, 118

Radioactive sources, 220

Radiography: with gamma rays, 219–20; with neutrons, 220–1; with X-rays, 217–19

Réaumur thermometer scale, 138

Refraction, 166; double, 239

Refractive index, 238

Refractometer, 238

Refractory materials, 143, 151

Refrigerator, working principles, 156, 157

Resistance, 97, 99

Rockwell hardness testing machine, 206

Röntgen, W. K., 216

Ruby, 239, *204*; manufacturing artificial, 152; synthetic, 242

266

Index

Dellow

Methods of science